W9-ALJ-413

Adding Parents to the Equation

Praise for *Adding Parents to the Equation*

"Let's face it: the way math is taught today is not the way most of us learned as students. That makes all of us feel uncomfortable and question whether it's a problem with us or a problem with how math is taught. If you're like me, though, you realize that while you might have been able to *do* math, you were more like a robot than someone who deeply understood what you were doing. That is why this book is so important for parents. Not only does it help you learn how to help your children, it tells you why this new approach is worthwhile and will help your children make sense of mathematics. Why do we carry the one or bring down the zero? Why do why we invert and multiply when dividing fractions? If you want to help your students deeply understand and appreciate mathematics, this is the book you've been looking for."

—Robert Kaplinsky, math consultant, www.robertkaplinsky.com

"At last! Hilary Kreisberg and Matthew Beyranevand provide an outstanding resource for parents, guardians, and all interested adults caring for elementary children. This resource enables adults to understand the whys and hows of learning and teaching mathematics today. Using a conversational tone, the authors offer clear explanations of how and when students acquire mathematical knowledge, explain and justify the approaches used in classrooms today, and present strategies and resources for parents to make mathematics a fundamental and foundational tool for all involved."

—Kit Norris, math consultant

"This is an essential book! Our culture promotes a paradoxical relationship with mathematics, to both revere and fear it, to be in awe of its power and to be intimidated by its strength. And for parents looking on, for whom the stakes seem high, this lycanthrope of a beast has just changed forms! *What is this 'new math'? What is my child doing in school? Why does everything look so unfamiliar? Surely 'math is math,' so what has happened here? Should I help my child take command of what I am seeing or protect her from it?* Kreisberg and Beyranevand address and answer these questions head on and do so with grace and the deepest respect. They eloquently bring sense to the *how* of mathematics as it is taught today as well as the *why* of it. Most importantly, they provide you full *empowerment* to understand and embrace the context and content of your child's mathematical journey. You are your child's most influential teacher, and this book is a gift to you. It will transform befuddlement and dismay into clarity and joy, and it will assist you in helping your child find confidence, facility, and delight throughout the curriculum. It is the gift that will open the door to one of mankind's most stunning of practical and intellectual achievements of all time: mathematics."

—James Tanton, PhD, mathematician-at-large, Mathematical Association of America, and founder, Global Math Project (theglobalmathproject.org)

"*Adding Parents to the Equation: Understanding Your Child's Elementary School Math* by Hilary Kreisberg and Matthew Beyranevand takes on the issue of helping parents and guardians understand the importance of teaching mathematics today so students have a deep understanding of the why, how, and when (conceptual understanding, procedural skills and fluency, and application) of mathematics. This book will help parents gain a better understanding of the current math lingo, the *why* behind strategies used in the classroom today, and ways to support students learning at home."

—John W. Staley, PhD, past president, National Council of Supervisors of Mathematics, and mathematics educator, Baltimore County public schools

Adding Parents to the Equation

Understanding Your Child's Elementary School Math

Hilary Kreisberg
Matthew L. Beyranevand

ROWMAN & LITTLEFIELD
Lanham • Boulder • New York • London

Published by Rowman & Littlefield
An imprint of The Rowman & Littlefield Publishing Group, Inc.
4501 Forbes Boulevard, Suite 200, Lanham, Maryland 20706
www.rowman.com

6 Tinworth Street, London SE11 5AL

British Library Cataloguing in Publication Information Available

Library of Congress Cataloging-in-Publication Data
Names: Kreisberg, Hilary, 1988–, author. | Beyranevand, Matthew L., 1977–, author.
Title: Adding parents to the equation : understanding your child's elementary school math / Hilary Kreisberg, Matthew L. Beyranevand.
Description: Lanham : Rowman & Littlefield, [2019] | Includes bibliographical references.
Identifiers: LCCN 2018040346 (print) | LCCN 2018055544 (ebook) | ISBN 9781475833591 (Electronic) | ISBN 9781475833577 (cloth : alk. paper)
Subjects: LCSH: Mathematics—Study and teaching (Elementary) | Education—Parent participation.
Classification: LCC QA135.6 (ebook) | LCC QA135.6 .K74 2019 (print) | DDC 372.7—dc23
LC record available at https://lccn.loc.gov/2018040346

∞™ The paper used in this publication meets the minimum requirements of American National Standard for Information Sciences—Permanence of Paper for Printed Library Materials, ANSI/NISO Z39.48–1992.

Printed in the United States of America

Contents

Acknowledgments

This book would never have happened without Matthew. I am so thankful that Twitter connected us. I appreciate you trusting me throughout this process and am blessed that we have become closest friends.

To my amazing husband, Robert, thank you for understanding my need to work nightly and on weekends on this project for the past two years, undoubtedly losing out on important date nights and family time. To my mom, dad, and sister (Cindy, Jay, and Mandy), thank you for being my number one fans. To Ralph, my father-in-law, thank you for reading through the entire book and offering helpful feedback.

To James Tanton, one of my idols, mentors, and friends, thank you for your most thoughtful and detailed feedback on each and every chapter. You took on a great load of work in reviewing this book for us. And finally, thank you, Brittany Gonio, for your support, editing, and dedication to the Center for Mathematics Achievement at Lesley University.

—Hilary

Thank you to my parents for building a love of mathematics in me from a very early age. I hope that this love is now translating to my three amazing children: Emily, Alexander, and Nicholas. I would not be able to accomplish anything without the love and support from my wife, Valerie. Thank you, baby.

Much appreciation to James Tanton for your guidance and friendship. I am in awe of you as a person and mathematician. Finally, Hilary, you are a rock! Thank you for pushing this project to the finish line and sharing many laughs in the process.

—Matthew

Introduction

> "I don't know that way. Why would they change math? Math is math. MATH IS MATH!"
>
> —*Incredibles 2*

It doesn't matter where we are—the supermarket, on vacation with our families, or working out at the gym—we find ourselves constantly overhearing frustrated parents complaining about two things: their inability to help their children with their math homework and the way math is taught today in comparison to when they learned it. After engaging with hundreds of upset, worried, confused, and angry parents, we realized that our nation faced a critical issue: the home-to-school connection has broken.

If we want our children to succeed academically, then we have to ensure that parental involvement remains intact. As important as teachers are to the education of our students, parents are a child's first and primary teacher. We could not do our jobs without their support. Therefore, it is our intent for this book to begin to repair the relationship between schools and the larger community by providing a rationale for why children are learning math differently today than their parents did when they were kids and offering resources to use at home to support these changes.

At its core, this book is written for parents or guardians of elementary-aged children but can also be appreciated by parents of middle or high schoolers as well, as we draw connections beyond elementary mathematics. While the book was written primarily for parents, teachers and those who are curious about math education today will also enjoy this read and learn something new. As we wrote this book, we kept in mind three goals that we wanted to accomplish:

1. Describe *how* children are learning mathematics today.
2. Explain and justify *why* learning mathematics through this approach is beneficial.
3. Advise *how* to help children succeed by supporting the math they are learning at home.

These goals were determined from an informal survey we conducted in October 2017 when we posted a questionnaire on the internet and asked parents from all over the nation to respond. Of the parents who responded, *60 percent* said their school has not provided parent information nights about modern math instruction. This statistic supports the common explanations we received that describe how parents feel while helping their children with their math homework. Look at table I.1 to see the four most common words parents used and generalized explanations that were developed directly from the most common responses.

For almost a decade, parents have been feeling frustrated, confused, worried, and intimidated as their children come home using a plethora of drawings and unfamiliar strategies to solve problems, some problems so foreign that parents say they feel as if they are speaking a different language from their children. For years, we have been asked by peers, colleagues, and family friends to write a book sharing our understanding of math education today with the hope that we can inform the larger community.

When the Common Core State Standards were released and adopted by states in 2011, the implementation and rollout left parents confused about what their children were learning and teachers in need of professional development to be able to teach what was expected of them. Years later, parents are still confused. We empathize with you as parents and realize that your beliefs about what you know about math have been challenged. This has left you defenseless and feeling intimidated, frustrated, worried, and confused.

Table I.1.

How Parents Feel	*Explanation*
Intimidated	Because their children are learning math in a totally different way, parents don't believe they can be helpful.
Frustrated	Parents feel unintelligent when unable to do "third-grade math homework."
Worried	Parents are anxious that their children will fail because of them.
Confused	Because parents rarely see anything familiar, math feels like a foreign language.

Before you read on, we ask that you be willing to accept and read what we offer without being influenced by what you *want* or *don't want* to be true. For example, if you read this book through an angry or refusal-to-change-your-thinking lens, then you won't absorb all the deep knowledge we are sharing that will help you better communicate with your child.

It is our intention to reduce the confusion, ease the stress, and arm parents with knowledge and strategies to be able to talk mathematically with their children. We hope this book helps parents feel more comfortable, confident, and aware of the way elementary math is being taught today in American public schools, ultimately leading to a shift in the culture and vision of math instruction.

The chapters of this book are ordered in a learning progression. Therefore, we encourage parents to read the book from start to finish, rather than reading sections out of order. Building the foundation solidifies a deep understanding and true number sense that enables one to move on to learning about operating with numbers.

Chapter 1 is designed to help parents develop a solid understanding of *why* children today are learning math differently than how they learned it. Equally as important as the understanding of *why* is the awareness of *how* parents can influence their child's love or hatred of math.

In chapter 2, we draw from top studies on mathematical mindsets to break down how to be the most effective support system for a child. Ensuring that you foster a growth, optimistic mindset will assist in your development of new skills.

Chapter 3 focuses on early numeracy and the building blocks of number sense. In order to develop a deeper understanding of the operations with numbers, you will first need to solidify your understanding of numbers themselves.

Chapter 4 came about as the result of frequent complaints from parents. Parents are constantly venting to us that they don't know the new math "lingo." In addition to a glossary, we felt it necessary to devote an entire chapter to breaking down terms that teachers or children may use around parents and detailing various mathematical tools used in classrooms today.

Chapters 5, 6, and 7 are meant to give parents a deep look into operating (adding, subtracting, multiplying, and dividing) with whole numbers and fractions.

To help parents see the value in the work elementary-aged children are doing, we wrote chapter 8 specifically to outline how middle school math builds on the math children learn in elementary school and how some of the same tools and strategies can be applied later.

In chapter 9, we share strategies, tips, and games parents can use at home. Finally, chapter 10 concludes the book with some final thoughts.

Throughout the book, parents will see tip boxes that integrate research and suggestions relevant to the context of the chapter. Parents will also see in the tip boxes examples of mathematics from ancient cultures, such as Babylonians, Egyptians, and others, demonstrating that the strategies children are learning today are not "new." There is also a glossary at the end of the book, and terms included in the glossary are bold in the chapters. Feel free to use this glossary as you read to help you make sense of the language.

We, as authors, realize much of the math we know today is a direct result of being educators—we have been immersed in deepening others' understandings of mathematics, and in order to do that successfully, we had to develop a better understanding ourselves. We contend that if parents develop a better understanding of the way math is instructed today, then they will be better prepared to support their children's education, eventually leading to increased student academic achievement. We hope this book answers many, if not all, of your questions and concerns, and we encourage you to discuss your learnings with your fellow parents.

Chapter 1

Why Is Math Taught Differently Than When I Learned It?

"The mathematics students need to learn today is not the same mathematics that their parents and grandparents needed to learn. When today's students become adults, they will face new demands for mathematical proficiency that school mathematics should attempt to anticipate."

—National Research Council (2016, p. 1)

Change is difficult for many people and even more so when the change seemingly occurs without forewarning and appropriate resources to accommodate. We want to make something clear: the manner in which our nation shifted the teaching of mathematics was done without consideration to the most important stakeholders—namely, parents. While it appeared as if it were an overnight change to many, the Common Core Math era, as we call it, had actually been researched for decades by experts in both the mathematics and education fields. Unfortunately, in most cases, notice to parents only came through talking to their children, leaving many frustrated as they began to realize they weren't able to help their children in math any longer.

There are many reasons why math instruction today differs from the way many of us learned. One vision is that society has become increasingly globalized and interconnected; we live in an information and technology-based world that is rapidly evolving. Due to this, basic content knowledge is easily accessible through computers and devices. Given that most information is readily available at the blink of an eye, human beings need to be better versed in complex problem solving and decision making, as those are skills computers and devices cannot yet take away from humans.

In terms of math instruction, this means that knowing 5×6 is 30 is not enough. A calculator, computer, or device knows this, too. But what a device cannot do is make sense of and reason through problems or make informed decisions that are not directed by some type of formula or algorithm.

We are living in a time period of major transformation, referred to by economists as the Fourth Industrial Revolution. Ensuring that our children are receiving an education that prepares them for a world of artificial intelligence and automation is key. According to the 2016 Future of Jobs Report put out by the World Economic Forum, "By one popular estimate 65% of children entering primary schools today will ultimately work in new job types and functions that currently don't yet exist" (p. 32). In the 2018 Future of Jobs Report, the World Economic Forum predicted that, by 2022, "human" skills such as "creativity, originality, initiative, critical thinking, persuasion, and negotiation" will remain crucial or even increase in demand, as will "attention to detail, resilience, flexibility, and complex problem-solving" (p. 12). For the education system, this means redesigning the learning platform to mirror what students will eventually meet when they are ready for college or a career.

The traditional ways of learning no longer prepare students for the future that lies ahead of them. The 2018 World Economic Forum concluded that there is an urgent need to improve education and skills levels, in particular in STEM (science, technology, engineering, and mathematics). The new impacts of updated technology remind us that our children need to be prepared to enter a workforce that will rely heavily on the human abilities to think, problem-solve, and reason. It is our job as educators to prepare students not just for today but for their future, so we must teach mathematics that will be applicable and useful in years to come.

In the past, students were taught that being *efficient thinkers* makes a student good in math. Efficient thinkers believe that math should be performed by following steps and procedures as quickly as possible with the least amount of resources, effort, and time invested. This is reinforced with traditional algorithms once learned in school, where many adults claim to not have understood what they were doing when they were "carrying," "borrowing," or "putting a zero" but did understand that it was fast and got them to the right answer (unless they were careless). In fact, in the past, someone who could get an answer fast was considered good at mathematics.

As you will see in chapter 2, study after study has shown that by placing time constrictions on mathematical thinking, we restrict students' abilities to show true understanding and can even contribute to added anxiety within students around math (Boaler, 2014). Between the studies conducted around math achievement and the enhancement and improvement of technology, we face new ways of thinking with which we need our mathematics instruction to reflect.

Students today are expected to become both effective *and* efficient thinkers. Effective thinkers are taught to think long term. This may mean using what appear to be complicated or longer methods for solving problems and perhaps not getting to the answer at first. This process allows students to develop a foundational understanding of important mathematical concepts that will eventually enable them to compute with facility.

To be an effective thinker, one must also question, critique, and reason, not just get an answer. Effective thinkers know how to be able to apply their understandings in alternative ways. Coupled with the ability to be efficient, students will be armed for success when facing unfamiliar problems. These types of thinkers are innovative and can quickly adapt their strategies, if needed, to meet their intended goal.

Essentially, school mathematics today prepares children to be creative problem solvers who find multiple solution paths. In other words, students are no longer expected to just get answers, but to focus on the process. Students are being taught to notice problem structures in different contexts and use their reasoning skills to make sense of problems. Think about the skills needed for twenty-first-century learners. Which do you think will better prepare students for using those skills that are listed? Teaching them to be *efficient* or *effective* thinkers?

In addition, each decade a plethora of mathematics education studies detail new and updated findings showing how children best learn mathematics. It is widely known today among educators that students better understand mathematics when taught through guided discovery, a teaching method where the teacher provides students support, but students ultimately develop the math rules themselves, as opposed to learning them through memorization, as many of us experienced.

In essence, the role of the teacher has shifted, as educators are no longer seen as the holders of the knowledge. Children today are given chances to explore, ask questions, and be curious thinkers, who are then led by highly qualified instructors toward connecting their inquiries to the intended mathematical results. This then leads students to being able to generalize, or conclude, rules that seem to always work.

We cannot express how many times we have witnessed parents, specifically parents who are frustrated helping their children with homework, end up just *telling* the student a rule that they learned in school and asking the student to use the rule to get the answer. This mentality, which will be further explored in chapter 2, takes away from the deep learning we are working so hard to give children.

Scientific research has shown that when students do work correctly, their brain does not grow; alternatively, when students struggle or make a mistake and *learn from it*, their brains grow and they make connections (Moser, Schroder, Heeter, Moran, & Lee, 2011). Look at the two scenarios in figure 1.1: (1) where a parent gives the answer to a question, and (2) where a parent

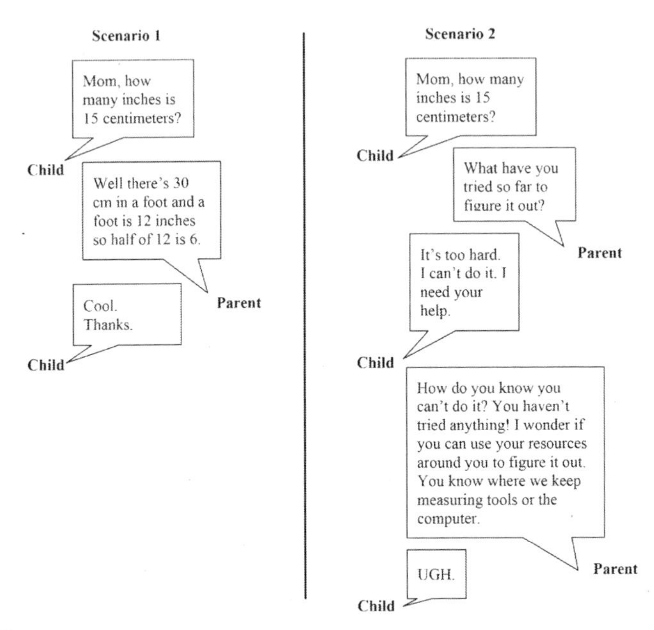

Figure 1.1.

allows a child to struggle, make a mistake, and learn from it. As you look at the conversations, think about how scenario 2 is more beneficial to the student than scenario 1.

So, what do you think? In scenario 1, the child gets the answer; there's no fighting, crying, or frustration; and the homework is completed. Why aren't we satisfied with that? That sounds like a parent's dream! Well, tomorrow, when that child has to do a different problem in math class that is similar, what strategies will they have to succeed other than learned helplessness? They will most likely raise their hand before they've even attempted the problem and ask for help. The teacher will most likely follow up with scenario 2 because teachers no longer are the people who hold the knowledge—our roles have shifted, as too have parents'. Scenario 2 better prepares your child for success, even if it means that, in the beginning, there's crying, screaming, frustration, and maybe even incomplete homework.

In scenario 2, the parent should give the child some space. Often, when given space, but held to a strict and structured environment, the child will persevere, or try to succeed, once they realize the parent isn't actually going to help. *Rule #1: Say what you mean and mean what you say!*

In this case, a child might go find a ruler, or look up "centimeter to inch conversion" on the internet. This resourcefulness is the beginning of problem solving, and they did it with minimal guidance. Tomorrow in class, that child will be more resourceful and independent. This is why it is important to let your child struggle. We promise you, it won't hurt them!

Should the child not attempt anything after five minutes, then the parent may want to revisit behavioral and homework time expectations in the home or investigate what's causing the issue. Oftentimes in this situation, it's a behavioral issue rather than a math issue.

Although mathematics instruction today is different from when most parents attended elementary school, these ideas are anything but new. The math today does not differ much from the way ancient mathematicians did calculations. During the days of the Ancient Egyptians (2200–1600 BCE), Babylonians (2000–600 BCE), and Greeks (600–100 BCE), mathematics was based on conceptual understanding, not memorizing procedures. **Conceptual understanding**, a phrase you will read often throughout this book, just means that students know more than basic facts and procedures to get to answers.

Did You Know?

Doctors used to perform surgical procedures called lobotomies, which they attributed to curing mental illness. During this practice, doctors either drilled a hole in the skull to remove part of the brain or would insert an ice pick through the corner of the eye to puncture the brain. At the time, this was considered a medical breakthrough, and the inventor was even given a Nobel Prize. Over time, better research and deeper understanding led the medical community to shift their practice. Similarly, in math education, instruction shifts as better research becomes available.

As noted earlier, as our society continues to become more technology driven, we need to identify ways in which our students' knowledge cannot be replaced by computers. Computers follow rules and algorithms; humans should understand how those algorithms work and be able to adapt them to suit other needs. Our methods of teaching math today encourage students to use mental math to make sense of the problems they are solving, motivate them to use tools to develop a deep understanding of problem solving, and support their communication of the math to others.

THE PEDAGOGY OF MATHEMATICS

Think back to when your child was first born. What was your newborn capable of doing (besides sleeping, eating, and pooping)? How did they learn?

Well, they started by touching. They touched (and tasted) everything! The sense of touch offers informative cues about objects around us and allows our brains to make important neural connections to experiences. As infants become toddlers, touch is still incredibly important for beginning experiences, but then we start using books and pictures of objects they have seen to help build associations and connections. Eventually, we strip away all pictures and teach them to read words without pictures.

Most math is taught today through a three-step approach, very similar to the way our brains learned as we progressed from infants to toddlers and on. In mathematics, this instructional sequence is best known as the Concrete-Representational-Abstract (CRA) approach. Just as the name suggests, in this method children begin by learning things in a hands-on manner, then drawing pictures or representations of the hands-on work, and then finally translating the representations into symbols (math). Below we provide more details on each of the three stages.

Concrete

The **concrete** stage is also known as the "doing" stage. If you are learning to add two numbers, then you should first start by physically *joining* or *putting together* two items to see what happens. In a classroom, this might look like a child connecting two blue cubes to one red cube to recognize that they are holding three cubes altogether (see figure 1.2).

Representational

The **representational** stage is also known as the "seeing" stage. Here, it is vital for children to make associations and connections between what they have physically touched to a drawing or representation of some sort to develop a deeper understanding.

In our previous example, after students have had enough practice connecting cubes, or "doing the math," they need to "see" and reflect upon what they just did. During this stage, we would then have students draw a representation, or picture, of what they were just modeling. It is key to note that students

Figure 1.2.

Figure 1.3.

are just drawing a *representation*; for our cube example, the representation does not need to look *exactly* like the three cubes to show the important information needed. Figure 1.3 shows how a child might draw the three cubes.

Essentially, this stage encourages students to see a common structure between different scenarios (in this case 2 blue cubes and 1 red cube, but after a few other similar cases, students notice that they can represent 2 green apples and 1 red apple with the same picture to show three of something). In order to make sense of something, the brain must make a connection to what it already knows (concrete-representational connection).

Abstract

The final stage is the **abstract** stage, also known as the "symbolic" stage. Once students have used physical objects to solve a problem and have then recorded those objects in a pictorial way, students begin to learn how the problem is presented symbolically. Using our cube example, a student in the abstract stage would solve the problem by writing or thinking $2 + 1 = 3$. They no longer need to touch or see the cubes or pictures of the cubes to think mathematically. Learning to represent a situation symbolically is challenging, as it decontextualizes, or takes away the context. This is why it is called the "abstract" stage.

This sequence of learning allows children to inquire, investigate, and create conjectures or ideas that are then tested and refined through further development. By allowing children to develop their own generalizations, they feel more invested in the math. It is important that parents understand these three stages and their benefits, so as not to interrupt the learning trajectory of their child.

FORMAT OF THE BOOK

Now that you have a foundational understanding of how your child may be taught math and why math instruction is different today than when you were in school, we will equip you with as many tools as we can to help you best support your child's learning of math at home. Chapters 2 and 3 will help you better understand your own thinking of math and how you can enhance your mindset and language to support your child's mathematical development. Chapters 4–7 will give you a detailed overview of the mathematics itself so you can feel more confident with the strategies your children are learning and hopefully be able to talk with them about their math experiences, successes, and challenges. Chapter 8 will help you draw connections between the math your child is learning in elementary school to the math they will learn in middle and high school. Finally, chapters 9 and 10 summarize how to use all this new information at home.

Read on to see how you and your child can develop a love of mathematics together.

Chapter 2

Mindset

"If parents want to give their children a gift, the best thing they can do is to teach their children to love challenges, be intrigued by mistakes, enjoy effort, and keep on learning."

—Carol Dweck, professor of psychology, Stanford University

As a parent, helping your child understand the math that they are learning can be a truly arduous task, especially when the math looks drastically different than it did when you learned it. How are you expected to support your children's learning when you can't tell what they are doing?

While this book will show you in detail various strategies that are taught today, this chapter is by far the most important for you to absorb. As the parent, you are your child's first teacher. We encourage you to support your child's learning, but we also want to remind you that it is best if you leave the teaching of the math content up to the schoolteacher and use your time at home to develop the *habits of a mathematical thinker* instead. There is no doubt that family involvement is one of the keys to student academic success, but not all help is good help! Depending on the mindset of the parents, outcomes can be profoundly impactful or ineffective. It all comes down to both you and your child's mindset.

The term *mindset* has been coined by Stanford University professor Carol Dweck and refers to your internal beliefs and assumptions. Your mindset dictates how you handle situations in your daily life. There are two predominant opinions regarding learning: a *fixed mindset* and a *growth mindset.*

Before exploring the idea of mindset in depth, we would like you to take a moment to determine your own mindset through a short questionnaire. Please take a minute and answer the questions in table 2.1, as honestly as

Table 2.1. Mindset Questionnaire

	Strongly Agree	*Agree*	*Disagree*	*Strongly Disagree*
1. Your intelligence is something you are born with; you are either born intelligent or not.				
2. Everyone can learn to become musically talented.				
3. I get frustrated when someone tries to give me feedback they think is useful.				
4. If you can't sing when you're young, then you won't be able to sing when you're old.				
5. An IQ test determines your intelligence; that number won't ever change.				
6. Telling children they are smart helps them do better.				
7. The more effort that you put into something, the better at it you will be.				
8. If you are smart, then you do not need to put in a lot of effort to do well.				
9. Albert Einstein was considered a genius. I'll never be as smart as he was.				
10. You can always improve your intelligence; you are not born with a set intelligence.				

you can. Select the appropriate box based upon your level of agreement with each statement by placing a check in the column that identifies the extent to which you agree or disagree with the statements. The survey goal is not to be definitive, and your answers won't be either—they can always be qualified—but use this survey to get a general sense of where you might stand.

Table 2.2. Mindset Scoring Rubric

	Strongly Agree	*Agree*	*Disagree*	*Strongly Disagree*
1. Your intelligence is . . .	0	1	2	3
2. Everyone can learn . . .	3	2	1	0
3. I get frustrated when. . . .	0	1	2	3
4. If you can't sing when . . .	0	1	2	3
5. An IQ test determines . . .	0	1	2	3
6. Telling a child he or she . . .	0	1	2	3
7. The more effort that . . .	3	2	1	0
8. If you are smart . . .	0	1	2	3
9. Albert Einstein was . . .	0	1	2	3
10. You can always . . .	3	2	1	0

Table 2.3. Mindset Scores

Total Points	*Likely Mindset*
0–8	Strong Fixed Mindset
9–17	Fixed Mindset
18–23	Growth Mindset
24–30	Strong Growth Mindset

Then use the Mindset Scoring Rubric in table 2.2 to determine your score. This can be completed by finding the sum to determine a rough estimate of your mindset.

The higher your point total, the more likely it is that you have a growth mindset. If you scored very high (24+), then you are open to learning and growing and believe that, with effort, you can get "smarter" (see table 2.3).

If your score was in the single digits, then you have firm beliefs that intelligence is not something you can improve. A fixed mindset will prove to be challenging in parenting a young child to develop mathematical thinking habits. We caution you to read some scientific studies that may help you develop a growth mindset before continuing with the rest of the book.

FIXED MINDSET

A fixed mindset is the belief that one's intelligence is determined at birth and does not change. Students with a fixed mindset in mathematics are often heard saying comments such as, "I have never been good at math; my parents were bad at math and so am I. It doesn't matter how much I study, I just won't get it." They become very discouraged when they make mistakes and develop a defeatist attitude. They buy into the adage that you are "smart," "average," or "dumb."

Parents can also be heard using fixed mindsets in math, too. Too frequently we hear parents say, almost proudly, that they were not good in math, and as a result, it is "okay" and "a given" that their children will not be good at math, either. Other times, we hear kids ask a parent for help with their math homework, only for that parent to affirm, "I don't know how to do this 'new math.' " Kids hear what you are saying, and they internalize it. Wouldn't you much rather have them hear you promote that you are a learner and willing to put in the effort to grow your brain?

The mindset that we have determines the view that we take of the world and impacts all of our choices throughout life. Individuals with a fixed mindset look for someone or something to blame for their failures. For example, children with fixed mindsets might say things like, "The teacher picks on me . . . If you would have helped me then I would have gotten it . . . You made me mad and I couldn't study." Blaming others does not improve performance. In order to improve performance, one needs a growth mindset.

GROWTH MINDSET

In contrast to a fixed mindset, a growth mindset is the belief that your intelligence is incremental; in other words, you can continually improve it. Students with a growth mindset see school as a challenge—a good one. They believe that they can improve if they put in enough effort.

Students with a growth mindset in mathematics can be heard saying things such as, "I can't do this yet, but I will if I put in more effort. Mistakes help me learn. This problem is challenging, but I will persevere through it!" Every so often, we hear parents use a growth mindset when talking to their children. For example, rather than complimenting a child's intelligence, parents have said things such as, "You did a great job spending so much time working through that problem. That effort will pay off." The use of positive praise in relationship to effort, not intelligence, has shown great impact on student capabilities academically.

Did You Know?

Research shows children's beliefs about their own intelligence and ability to learn strongly impacts the choices that they make, such as the time and effort they put into their learning (Dweck, 2006).

People with a growth mindset know that success is the result of hard work and effort. They think reflectively on what they can do differently in order to be successful. It is important to understand that "growth" is not synonymous with earning all As in school. As most elementary schools and many middle schools move away from letter-based grading and to standards-based grading, the hope is that we can all change our definition of growth and success in mathematics.

Growth means pushing yourself beyond your current capabilities and always doing your best; grades do not determine that. We want our students and children to *want* to do better and be risk takers. Children should explore possibilities and dream big and then back up those dreams with the attributes necessary to becoming successful.

The bottom line? We live in a world in which it is socially acceptable for people to believe that being bad at math is genetic. Look at figure 2.1 for an example of what would happen if we replaced "math" with "thinking" in our normalized, socially acceptable language. You would never hear people chanting that they aren't "thinkers." You know you've heard many times people say, "I'm not a math person." Could you imagine if you heard someone say, "I'm just not a thinking person"? Why do we find it socially acceptable to be proud about our fear of mathematics, yet in similar fashion, it is not a norm to boast that you can't read or think?

It's time we adapt our thinking; as adults, we need to be careful of our language. Kids hear what we say, and because we are our child's first teacher, they

Figure 2.1.
Dyke, E., & Dyke, J. (2014). 10 things every parent should know about math. Queen Creek, AZ: J&E Dyke Enterprises, LLC.

believe what we say! More important, we have to model open-mindedness, a willingness to learn, grow, and develop.

No one is perfect—be sure to remind your children that you make mistakes, but that you use them as learning opportunities. For example, you may come home and say, "Oh man, I made a mistake at work today, but I got it fixed and I will remember next time what I need to do so I don't make that mistake again." Let them know that making a mistake is just part of the learning process. Instilling a love of learning within your children will help them become better problem solvers, ones who persevere and work through what they perceive as challenges.

Did You Know?

A study with middle school students looked at the impact of fixed versus growth mindsets on achievement in math—a subject that many students find challenging. Students with a growth mindset earned higher math grades over time compared to students with a fixed mindset (Blackwell, Trzesniewski, & Dweck, 2007).

Believe it or not, our brains act like plastic. Essentially, our brains grow and shrink depending upon how we take care of them. This neuroplasticity helps support the notion of growth mindset. The more you train your brain to be open to new ideas and grow, the better the chances that your brain will actually do that!

Michael Merzenich, author of *Soft-Wired: How the New Science of Brain Plasticity Can Change Your Life,* discusses 10 core principles that must occur for you to rewire your brain. We hone in on a few of his principles that we believe tie directly to the mindset parents and students need to have to become better at mathematics.

Change occurs when the brain is in the mood for it.

This means that we need to be focused on helping students to be in the right frame of mind in order to learn. We need to believe in a growth mindset and encourage our children to believe the same—their brains will grow if they believe they can improve. Parents have to provide an environment that values learning, first and foremost.

Encouraging your children to spend time at night "growing their brain" is a great way to instill these values. This could be as simple as establishing a time frame in your household where everyone reads, writes, or does a puzzle. In addition, parents have to talk the talk and walk the walk. Your beliefs have a powerful impact on your kids' beliefs. Have a growth mindset about developing a growth mindset!

Children who know that the brain can get "smarter" will do better in school because they become empowered to take charge of their own learning. They understand that failure is part of the learning process. Meanwhile, those with a fixed mindset tend to focus on how others perceive them and avoid situations in which they might have to work through a challenge and/or risk failure. The best way to get a child to develop a growth mindset is for you to model one.

The greater your effort, motivation, and awareness, the more your brain can change.

The more effort you put forth, even when the task is difficult, means that you are going to have positive outcomes in learning. Hard work does pay off! It is important that students are able to focus on the task at hand. Parents need to make sure that their children are in a place to be able to focus without distractions. The more challenging the task, the more important it is to eliminate distractions such as TV, music, and sometimes even interactions with friends.

It is also important to model for children that persistence, even when challenging, helps develop problem solvers. When children finally "get it," then it will remain with them for a longer period of time. The next time you are about to give up on something that is personally challenging for you, think about modeling perseverance for your children. For example, if you were planning to head to the gym but find excuses not to go, push yourself to change your mindset so your children can watch you work through a challenge.

Your brain needs to be trained to record change that is worthy of permanence.

Your brain is always observing what you are doing and taking notes! Your brain only registers a change as permanent if it experiences enough of the activity and realizes that the activity is meaningful and won't harm you. This means that initial changes are temporary. Your brain has to decide that the information is "important" in order for it to store information permanently. Therefore, parents should encourage children to persevere through challenges and take their time so the brain registers that shifting the mindset is important.

Every step they take to complete homework or study for a test instills a work ethic in them that will carry them through tough times. Preparing children for a successful future can be difficult (perhaps the assignments are not always fun or they might have experience struggling), but when they recognize the importance of what they are doing, their brain will take over and begin to automatically store the important information.

Your memories facilitate your learning.

Our brains are capable of incredible things. When you are learning a new skill, your brain knows to throw away the memories of failed attempts, while preserving the good attempts. This is why it is important to keep working at a problem until you have a successful attempt, even if it means taking a hiatus and coming back to the work another time. Having a growth mindset means toughing it out through the failures until you get to a success because that is when growth occurs.

We understand how tempting it is to just give an answer to a child when we see them frustrated and upset—no one wants to see their baby struggling—however, their brain is not recording and remembering the good attempts because there was no attempt. We encourage you to allow your children to work through struggles and, when necessary, supply hints or ask questions that can lead a student in a helpful direction.

We suggest asking questions rather than offering hints. By constantly asking questions, your child will begin to see that you will not be a source of answers. Eventually, they will stop relying on you. More than that, though, questioning is a technique that stimulates thinking, yet scaffolds, or guides, students to think differently. Look at the examples in figure 2.2 to see the difference between a parent who gives an answer and a parent who guides their child to learning.

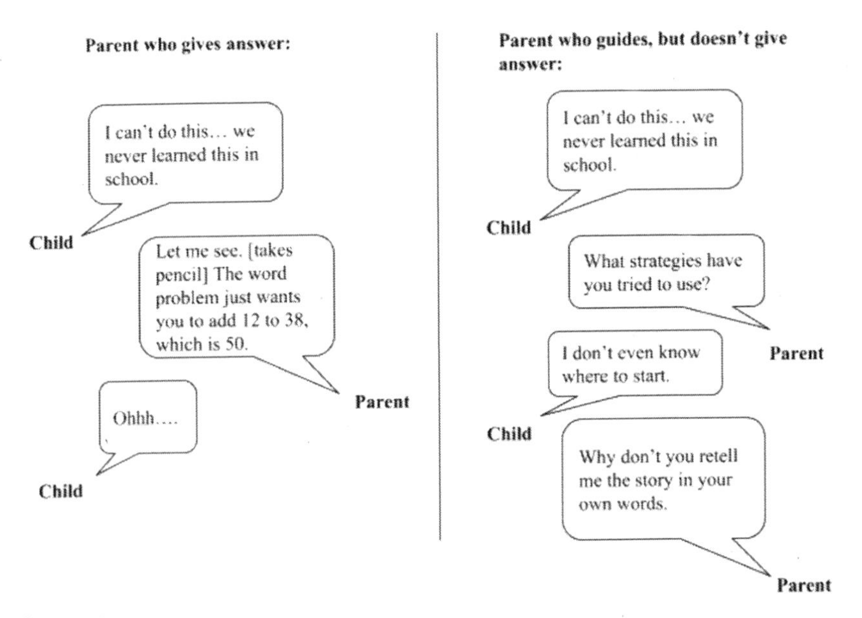

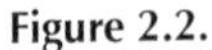
Figure 2.2.

Some examples of questions you can ask your child that may encourage them to persevere through a tough situation include

- "Can you explain or justify your thinking?"
- "Why does that work?"
- "Do you have any guesses as to what that word could mean?"
- "Do you see any patterns in your work?"
- "How is this problem similar to or different from the one you just did?"
- "Is your answer reasonable? Does it make sense?"

The more something is practiced, the more connections are changed. Once your children experience success, they will begin to make connections between ideas and gradually become more independent, persevering a little bit more each time.

It is just as easy to generate negative changes in your brain as it is positive ones.

This is where we have to really be careful. It is just as easy to create a permanent mindset of "I can't do" as it is to develop an "I can do" attitude. As noted earlier, try to avoid saying things such as, "I was bad at math; don't ask me!" as it perpetuates a fixed mindset. Use every opportunity to share with your child hurdles that you have had to overcome as well as areas where you have failed but persevered until you succeeded; continue to encourage them in all aspects of their life to be an "I can do" person.

Did You Know?

When family engagement is well structured, there is a direct association with higher grades, test scores, and motivation, along with improved attendance. Developing a growth mindset as part of your family values is one way to lead toward higher student achievement! (See Walker, 2014.)

THE ROLE OF PARENTS IN MATHEMATICS EDUCATION

When parents begin and maintain a good communication with the school system, they will see students grow in a multitude of ways, not just academically. When that partnership blossoms, it is easier to see if a child begins to stumble academically and/or socially. For most parents, children are in school longer each day than you get to spend with them, so being in a partnership with the school system will give you valuable information regarding the development of your child.

View the job of parenting as preparing your precious child to have the ability to move beyond high school and become an independent individual who has the ability to take care of themselves physically, mentally, socially, and financially. This process begins at a young age—yes, as early as elementary school—and you have to allow them to make choices and accept the consequences of their choices.

If they have waited until the last minute to complete homework or a project and they are going to have to accept a lower grade or a zero, then allow them to accept that consequence. *Do not do it for them or allow them to miss school.* Turn it into a teaching moment about the importance of managing time.

In order to achieve success, one must be able to overcome challenges, persevere, and problem solve. As a parent, you can help your child become independent and prepared for success by allowing them to make mistakes and learn from them, even when there are consequences. It might be better for them to go to school without doing their homework as opposed to you having to remind them multiple times every night or spending an entire night hand-holding them until they have finished.

Too often we see parents *doing* their child's math homework or turning homework time into hour-long painful sessions where the parent struggles to explain the mathematics to their child.

Picture this: your child comes home from school and starts his or her homework. It's only a matter of seconds before you hear "I don't get it" and you feel the need to step in and be the Homework Hero because you hate to see your kid struggle and, more important, you just want that homework finished so it doesn't take up your whole night. Sound familiar? How often have you played "teacher" at night to make sure your child goes to school the next day with a paper full of correct answers? For many, this is an all-too-common phenomenon. We are giving you permission to stop that and take those hours back.

It is much more advantageous to let your child problem solve, use resources (such as a notebook, a peer, or the internet), and actually struggle than for you to sit there helping them every step of the way. In fact, believe it or not, teachers prefer an empty homework sheet over receiving a perfect paper.

Should your child receive homework in grades K–2, and this happens, we encourage you to just attach a note onto the homework explaining what was causing a challenge (e.g., "she became overly frustrated and gave up"). If the child is in grades 3–5, we ask you to let the child write a note to the teacher explaining the struggle. This then gives the teacher a good starting place with the student, rather than seeing a perfect paper and perhaps letting a student fall under the radar.

Remember that there is a difference between parental *help* and parental *interference* when it comes to homework and any other school-related activity. Homework help means assistance with finding resources, providing a time and a place with few distractions, teaching time management, and encouraging students to develop study skills. Be sure to create an environment for learning.

When is the best time for your child after school to do homework? Think about the time frame from after school until bedtime. When are they at their best physically and mentally? Remove all distractions—particularly electronics—which can be troublesome in today's era where many homework assignments are digital. Make time for short breaks that allow the student to move around.

In addition, have daily learning discussions. The younger the child, the more specific the question has to be. If you ask, "What did you learn today?" that is entirely too broad for their little minds. Instead, ask:

- What did you learn in science today?
- Did you learn a new word today?
- How did the addition go in math class today?
- Did you make any mistakes today in your work?
- What do you need to remember so that you don't make the same mistake tomorrow?
- What do you have to try really hard at?
- Do you have homework?
- How do you feel about your ability to do the homework?
- What can I do to help you with your homework?

Remember, some of our greatest life lessons come from our failures. Nelson Mandela speaks to this when he said, "I never lose. Either I win or I learn." So does Thomas Edison when he proclaimed, "I haven't failed. I just found 10,000 ways that won't work."

Below are five famous individuals who used a growth mindset to overcome various setbacks:

Walt Disney. Did you know he was fired from a newspaper for "lacking imagination" and having "no original ideas"?

Albert Einstein. Did you know Albert Einstein wasn't able to speak until he was 4 years old? In fact, his teachers told his parents that he wouldn't amount to much!

Michael Jordan. Did you know that Michael Jordan didn't make his Varsity basketball team in high school?

Table 2.4. A Fixed Mindset about Learning Math

Child Says . . .	*Parental Response*
I can't do this.	"You either can't do it *yet*, or you *can* do this and need to find another strategy."
I'm not good at math.	"Math may not be your favorite subject right now, but you can learn to enjoy it."
I always do this wrong.	"Let's figure out where you are making the mistake so you can correct it and then practice doing it right."
I give up.	"Giving up is not an option! Think outside the box. What are some strategies you could use in order to help you understand the problem and/or the process?"
I will never be as smart as [friend/ classmate].	"Smart means different things to different people. What does [friend/classmate] do or say that you think makes them so smart? You show that you are smart when you [persevere; listen to others; tell the truth; etc.]."
This is too hard.	"It is hard, and I appreciate how you never give up. I know you are going to get this if you keep trying."

The Beatles. Did you know that the Beatles were rejected by a recording studio and were told they had no future in show business?

J. K. Rowling. Did you know that J. K. Rowling's manuscript of *Harry Potter* was rejected by 12 different publishing companies before finding success?

Parents—it is just as important that *you* have a growth mindset as it is for your child. If you are willing to learn with your child and model a growth mindset, then you are going to see your child make great strides in learning *and* prepare them to be independent individuals who will not be afraid to go out into the world and pursue their dreams. Table 2.4 contains a few examples of a student's fixed mindset about learning math and a parental response reflecting a growth mindset.

PARENTAL PRAISE

We cannot stress enough the dire need for parents to change the way that they praise their children. Instead of praising their ability or results, praise their effort that they put into mathematics or any activity. Praising their ability by saying things like "you are so smart" sets your child up for challenges later when they inevitably begin to struggle. When they have repeatedly been told that they are smart, they will likely lack the perseverance when math no longer comes easy to them.

Instead, praise your child in such ways as, "I am so happy that you took your time on that problem; you kept going and didn't quit!" or "Nice job putting in a lot of effort and showing all of your work on your test." Praise their effort, persistence, strategies, time management, and meeting of challenges. Do *not* praise their intellect ("You're so smart!").

If you find your child (or yourself) struggling to switch over from a fixed to growth mindset, start by encouraging your children (or yourself) to end phrases with "yet." Adding "yet" to negative phrases can instantly shift a person's mindset from *fixed* to *growth*. Children may not know their multiplication tables or how to divide with decimals *yet*, but they will one day! Before you respond to your child, be sure to reflect and think, "Will my reaction influence my child's future behavior? Am I encouraging learning, growth, and the acceptance of challenges?"

Did You Know?

Researchers conducted a study in which they observed how parents praised their children, ages 1 to 3. Five years later, the researchers measured the children's mindsets. They found that the more parents praised the process (effort, not intelligence) when their children were 1 to 3 years old, the more likely those children were to have a growth mindset five years later (Gunderson et al., 2013).

Another important step is to develop a collaborative relationship with your child's teacher. It is likely that you already have a positive relationship with your child's teacher related to behavior, assignments, and academic success. However, how much of the curriculum and instructional approaches do you know?

The teacher is going to be your primary source of information as to how much help your child actually needs from you. It is important that you follow the teacher's suggestions as to completing assignments and participating in intervention programs. If you can accept the constructive criticism that might come, which is designed to help you determine the areas in which your child needs to strengthen, it will take a lot of the sting out of it and allow you to make decisions that are in the best interest of your child.

Additionally, let your child struggle, but make sure it's productive. Children who *productively* struggle attempt to solve what may be a challenging problem, despite its difficulty, and look at their mistakes as learning aids. These children apply various strategies and usually find a solution. Children who are comfortable in this phase may put a problem aside, walk away, or take a break, but they are eager to come back to the problem at hand and attempt new strategies until a solution is identified.

On the other hand, students who enter the *destructive* struggle zone have run out of strategies and no longer feel equipped to solve the problem at hand. You will know when your child has begun to enter the destructive struggle zone when they become doubtful about their abilities to be able to be successful on the problem. Eventually, after enough self-doubt, the child will become apathetic, losing interest in the problem. Once a child doubts their ability to solve the problem and no longer cares to solve the problem, they may become frustrated. A child in this stage may feel abandoned and alone, which eventually leads to the student abandoning the problem or giving up. Review figure 2.3 to see what we deem the Five Stages of the Problem-Solving Struggle.

It is widely known and accepted that experiential learning is more effective than direct instruction (I talk; you listen). Students need to discover and learn on their own. In order to do that, they must feel comfortable testing out ideas, fixing mistakes based on results, and persevering, or putting in effort and not giving up. What we do not want is a child becoming frustrated and abandoning a problem. *If your child begins to show signs of frustration, then know that it is time to stop working and come back when your child is ready to make mistakes and learn from them. This is difficult but necessary.*

Encourage taking risks, failing, and learning from mistakes. Failure teaches great life lessons, particularly resiliency. No one wants to watch their own kids fail or struggle, but humans need to learn coping strategies for the emotions associated with struggle.

At some point in life, we are going to fail (most likely many times), causing us to feel sad, frustrated, and upset. Powering through the failures and the sadness is how we experience success. Homework is a simple place to

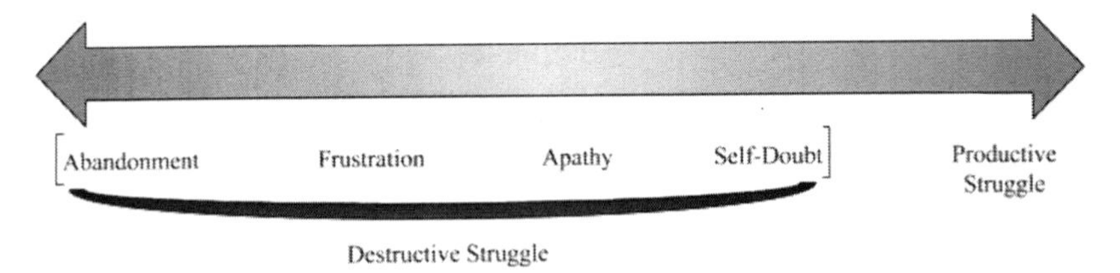

Figure 2.3.

start learning that process and help children find strategies to embrace a productive struggle. For example, let your child hand in homework that is incomplete or includes mistakes. This will help them learn, and their teachers would prefer it.

All in all, mathematics and mindset go hand-in-hand. Jo Boaler, a professor of mathematics education at Stanford University, has conducted multiple studies that show the following four results:

Being good at math is not a genetic trait. You can't be "bad" at math because your parent was bad at math. It also isn't a label for you. There's no such thing as a "math person" or "not a math person." *Everyone* can think mathematically.

Math is not about speed. Being quick to calculate does not indicate great math skills. Math is about reasoning, proving, and recognizing relationships.

Memorization in mathematics results in lowest achieving scores. Memorizing the multiplication table is not as helpful to understanding mathematics as recognizing patterns within the multiplication table and applying it elsewhere.

Math is a subject for *learning*, not *doing*. Math is not meant to be a performance but, rather, an area where people are engaged in thinking about and with the subject.

HOMEWORK HELP

We are always asked by parents for tips on how to help their children with homework. We strongly believe that by focusing on developing a growth mindset, parents will feel less dreadful of homework time, mainly because they will let their child struggle, make mistakes, and in turn, save themselves from the battle. The Homework Tips chart in table 2.5 offers several homework habits that you can develop in your home, in addition to fostering a

Table 2.5. Homework Tips Chart

Don't	*Instead . . .*
Send your child into a closed room with no supervision at a time of their choice.	Provide a "learning area" where you can supervise and where supplies are handy and distractions are minimal. Enforce rules and expectations for when homework is to be done and what quality is expected. Help children to structure time, space, and materials for completing homework. Be available.

(Continued)

Table 2.5. (Continued)

Don't	*Instead . . .*
	It is important that you choose times that you know are good physically and mentally for the child. For instance, right after school may not be a good time if they are either too tired or too hyped up right after school. Look for the optimum times and utilize them.
Reward or compliment your child's "smartness."	Reinforce your child's work by praising their *efforts*, not complimenting their ability to be correct. Go lightly on the rewards—students need to understand that homework is expected of them, and the reward is the learning and reinforcement of basic skills, not a tangible item. Rewarding with prizes can be a costly enterprise that sometimes can result in reinforcement of the wrong behavior.
Do the work for them.	Encourage them to look at their notes, textbooks, or online resources to find a strategy for doing it on their own. If they want to ask a friend how to do the process, this is okay, but remind them that getting an "answer" does not mean they have learned anything. Assist, help, tutor, but *never* "do" the work for them. Should your child struggle with homework, we suggest encouraging your youngster to write down the questions they have on a sticky note and approach their teacher the next day before class to find the answers. Teachers often receive homework from students with parent writing, notes, and comments. This does not support a child's independence and advocacy for their own work. It also doesn't tell the teacher *what* caused the struggle. Usually it's just a complaint that the child "couldn't do it," which is not very helpful for a teacher.
Engage in negative self-talk.	If you tell your child that you were never good at something, they will latch on to this as an excuse for them not doing well also. Instead, praise them for continuing to work on something even when it is hard. Remember, praise their *efforts and behaviors*, not their intelligence.
Tell your child "I don't understand how your teacher did it, so I am going to show you how I learned to do it."	If you do not understand the process that the teacher is showing, seek out resources until you feel confident that you can help them. Do *not* show them to do it a different way. Although it is hard to see, there is a bigger picture in which students are learning a process that is going to continue to be built upon from early elementary through the completion of high school. When they "miss" the process and just get a correct answer, they are creating gaps that will be hard to fill in later when they are asked to extend a concept and build on top of it. You have to help them develop an attitude that getting an answer is not "learning." Getting an answer only leads to short-term memory retention and the inability to apply the process to a multitude of situations. Even when a teacher's process doesn't seem to make sense, take a deep breath and realize there is a method behind the madness and encourage your child to persevere or keep trying.

Don't	*Instead . . .*
Refuse to help at all.	There are many ways you can help your child that still puts the burden of learning on them. Listen to them read aloud, be their study partner with flashcards or study guides. Ask specific questions about what they are learning. Ask them to summarize a book, a plot, or tell you what they know about a character in a book. Call out spelling words.
Consider successful completion of homework as every problem finished and all correct answers.	Remember that teachers use student homework (and in-class work) to determine how much more they need to teach of a particular subject. If you do your child's homework or don't allow them to show wrong answers, then they won't get the reteaching they need.

growth mindset. Use these tips to help you become positively involved in your child's academic life, without becoming *too* involved.

SUMMARY

In our work with parents, we have found that there are three main mindset shifts parents still need to make to foster a better mindset at home:

Math is not about getting the right answer. Shift from seeing math as *answer getting* to *problem solving*. Focus on your child's effort for solving problems, not whether they get the answers right or wrong or whether they earned an A or a C. Mathematicians do not prove their theories instantly; rather, they make a lot of mistakes and try various strategies until they can sufficiently justify one.

Being "stuck" or "challenged" is a good thing. Encourage your child to struggle, but make sure it is a productive struggle, otherwise your efforts will be fruitless. Parents tend to intervene too quickly when their child is challenged by a problem. If it were easy and straightforward, then why would it be called a *problem*? If we want students to be successful, then we have to give them opportunities to prove that they can apply themselves. You've heard the phrase "learn from your mistakes"—we learn best from trying something on our own, making a mistake, and fixing an error based on what we learned.

Praise the effort and behavior, not the intelligence. Telling your children that they are smart does not give them critical areas on which to improve. How will they know what they did well, other than by getting a right answer? What will happen in the future if they do not get a right answer? Will they be equipped with memories of strategies or compliments? Be sure you let your child know that you are proud, but you should not be proud because they are "so smart" but, rather, because they stuck with a tough problem or made a difficult decision.

Chapter 3

Early Numeracy

"Kids can rattle off their numbers early, often from 1 to 10, and parents are surprised and impressed. But it's a list with no meaning. When you say, 'Give me three fish,' they give you a handful."

—Susan Levine, Professor at University of Chicago

Many parents of children under the age of 6 wonder how much their child should know and what they can do at home to help their child prepare for formal schooling. Research shows that future success in mathematics stems from an early foundation of mathematical proficiency (Frye et al., 2013). Therefore, it is critical that families work with their children from an early age to develop a love of numbers and their connection to the real world. It is equally important that parents have a general understanding of how children develop *number sense*, or a flexibility with numbers, so as not to push their children beyond their capabilities. Number sense is the foundation of mathematics—it's what gives numbers meaning!

In this chapter, you will learn about how language impacts the learning of numbers and how one comes to make sense of numbers through a trajectory, or learning progression.

OUR LANGUAGE IS A BARRIER

Students who learn to speak Asian languages perform better in math early on than students who learn to speak English as a first language (Clements & Sarama, 2014; Sousa, 2015). This isn't because kids in Asian countries are better at math but, rather, because Asian languages require fewer words to know when counting. Did you know that students who learn English must memorize 28 words to be able to count from 1 to 100, whereas students who learn Chinese, for example, only need to know 11 (Sousa, 2015)?

For comparison, look at the 28 number words in English that students need to learn and compare them to Japanese and Chinese (table 3.1). What repetition do you notice between the Chinese and Japanese numbers? Look at the number 10 in all three languages. Now, look at the rest of the numbers and see where you see *ten* or *shi* or *juu*. You'll note that, in English, you will only find *ten* once, but in Chinese and Japanese, you find *ten* 18 times.

Additionally, in Asian languages, number words match the place value of the numbers. For example, *twelve* in Japanese is *juu-ni*, which literally means "ten-two." This makes it easier for students to later learn that 10 + 2 = 12, whereas students who speak English cannot make a similar connection. To make things even more challenging for English-language speakers, think

Table 3.1. Twenty-Eight Number Words in English, Chinese, and Japanese

Number	English	Chinese	Japanese
1	one	yī	ichi
2	two	ér	ni
3	three	sān	san
4	four	si	shi
5	five	wu	go
6	six	liu	roku
7	seven	qi	shichi
8	eight	ba	hachi
9	nine	jui	kyu
10	ten	shi	juu
11	eleven	shi-yì	juu-ichi
12	twelve	shi-er	juu-ni
13	thirteen	shi-san	juu-san
14	fourteen	shi-si	juu-shi
15	fifteen	shi-wu	juu-go
16	sixteen	shi-liu	juu-roku
17	seventeen	shi-qi	juu-shichi
18	eighteen	shi-ba	juu-hachi
19	nineteen	shi-jui	juu-kyu
20	twenty	er-shi	ni-juu
30	thirty	san-shi	san-juu
40	forty	si-shi	shi-juu
50	fifty	wu-shi	go-juu
60	sixty	liu-shi	roku-juu
70	seventy	qi-shi	shichi-juu
80	eighty	ba-shi	hachi-juu
90	ninety	jui-shi	kyu-juu
100	one hundred	yì bǎi	hyaku

about numbers such as *eighteen*. Children need to learn that *teen* in this case means *ten*, and to make matters worse, the first thing a child hears when saying *eighteen* is *eight*, not *ten*. This makes it troublesome to learn that *eighteen* is a 10 and an 8 and often causes children to write the number 18 as 81.

Because language plays a vital role in the development of number sense, be sure to *talk* math with your child before working with them to make sense of numbers through actually counting objects. For example, count down from 10 to 1 when a meal is ready, or point out that there are eight sides on a stop sign and count them aloud. By exposing your child to the language of numbers early on, specifically in their first year of life, you will better prepare them for learning how to count objects later.

DEVELOPING NUMBER SENSE

Education researchers have outlined various learning progressions that show a trajectory of what foundational skills one needs before learning other skills. Learning progressions help teachers identify how to enhance or scaffold a skill, depending upon a child's current ability. The number sense trajectory we discuss in this chapter shows how students learn to make sense of numbers.

Learning to make sense of numbers is complicated and not so black and white. We do our best to break down for you a general process showing how children learn about numbers, but do not worry if your child is showing signs of more than one stage at once. Each child is different, and while many follow this learning progression linearly, it doesn't mean that all do. Additionally, there are many other facets that contribute to this trajectory, but we have simplified the research into a bite-sized chunk for someone who may not be an educational professional. We want this to be helpful, not more confusing!

What matters most to us is that you generate a foundational understanding of how children generally develop number sense, so you allow your child to learn according to their developmental ability. We often see parents pushing children beyond their developmental capacity, solely focusing on how their child performs in math compared to other students, or to label the child "intelligent." As we saw in chapter 2, mindset is a major contributor to our incremental intelligence, not pushing students beyond their capabilities. Keep this in mind as you read this chapter.

As children begin to make sense of numbers, they will start to build connections between three main components that make up numbers: words (number names), quantities (amounts that make up a number), and numerals (symbols). Figure 3.1 shows the three aspects that make up a number. A student must internalize all three components to begin to understand how numbers are flexible.

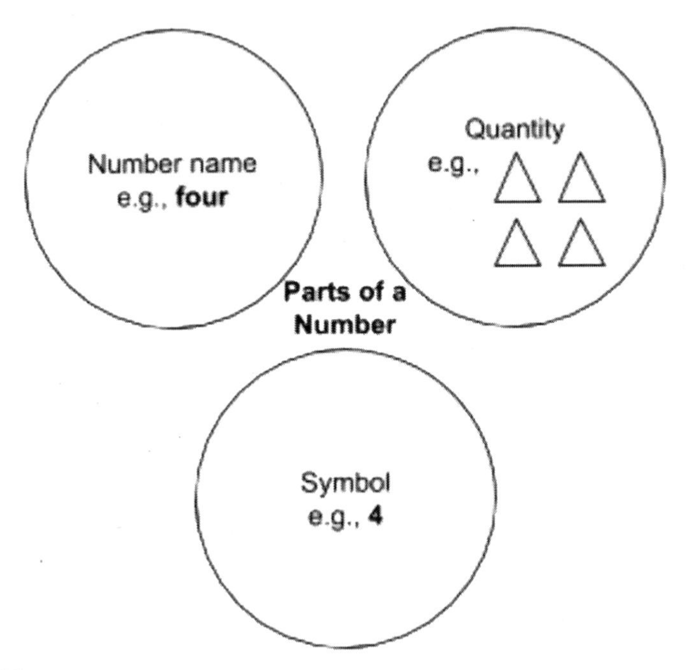

Figure 3.1.

This is the beginning of numeracy, or the understanding of what each number represents (think literacy for math!). Parents tend to get overly excited when a child can count to 10 or say the alphabet, but neither of these skills is useful if the child does not see a *relationship* between the counting sequence and quantities or the alphabet and words. Research has shown that children tend to make sense of numbers through a trajectory, or progression (Clements & Sarama, 2014; Van de Walle, Karp, & Bay-Williams, 2017). This developmental sequence is not a way of teaching, nor is it a pathway that can be mastered in a short amount of time. Instead, it is a summary of research showing *how* children typically learn to make sense of numbers. This process takes years to develop, so do not try to push your child to become proficient at it all immediately. We include in the overview of the trajectory below the average age children develop these skills; *however, it is critical that you realize that* average *does not mean "all" and your child may be ahead or behind these stages and that is okay.*

Before children learn to make sense of numbers, they start by simply noticing collections of objects. This is what researchers call **numerosity** (e.g., animals and humans are born with an innate ability to recognize that two quantities may be different). This targets the component of quantity and attaches meaning (number words) to those quantities. This stage usually occurs between ages 0 and 1 (Clements & Sarama, 2014).

Next, children learn to recognize quantities by **subitizing** (pronounced soo-bih-tize-ing), or knowing visual quantities (usually six or fewer) without counting (e.g., seeing five dots on dice and just "knowing" there are five dots). This stage progresses by cognitive ability, or the readiness of the brain. Children typically learn to subitize quantities of one, two, and three within the first two years of their lives and then can subitize quantities of four and five around ages 4 or 5 (Clements & Sarama, 2014).

After, children compare quantities *visually* (e.g., they can see that *four* apples is more than *two* apples). At this stage, children do not write inequalities such as 4 > 2. They may be able to write the numerical symbols for the number of apples, but it is more important that they are developing the language and associating *more* and *fewer* with visual quantities. Children at this stage are usually ages 4 to 5 (Clements & Sarama, 2014).

Then, children learn to say the *counting sequence*, or counting verbally starting from 1 (e.g., 1, 2, 3 . . .), but meaning will not be attached to this step until they've reached the next stage. As students verbally count, they begin connecting number words to numerals and developing an understanding of the three parts of a number. Most children by the age of 5 have reached this stage (Clements & Sarama, 2014).

After learning the counting sequence, children then begin to attach meaning to the numbers through **one-to-one correspondence**, or the ability to connect one number word to a quantity through counting (e.g., one object is assigned the value of one, but the second object in a counting sequence is assigned a value of two). This stage usually begins occurring around 3 years of age, but only typically for small groups of objects laid in a line. The full stage is typically achieved before a child turns 6 (Clements & Sarama, 2014).

At this point, children are ready to learn **cardinality**, recognizing that as they count, the last number they say represents the total objects in a set (e.g., 1, 2, 3, 4—*four* is the number in the set). This stage usually occurs between ages 4 to 5 (Clements & Sarama, 2014).

Once children have made a deeper connection to the counting sequence, then they develop *hierarchical inclusion*, or the idea that numbers are nested inside of other numbers (e.g., 10 is one more than 9, so 9 could never be 10, but 10 always has 9 within it). This stage usually occurs between ages 5 to 6 (Clements & Sarama, 2014).

The final stage of the number sense trajectory is *number conservation*, which is when children develop the understanding that a certain quantity can be reorganized and preserve its identity (e.g., five objects can be rearranged into groups of two and three, or four and one, and there will still be five objects). This stage usually occurs around 7 years of age (Clements & Sarama, 2014).

Let's look at these stages in detail, and as you read the stages, think about where your child may fall on this trajectory. *Please note that children do not need to know the terms used in the trajectory* (e.g., numerosity, subitize, hierarchical inclusion); *we simply use them so you can communicate mathematically with your child's teacher.*

NUMEROSITY

Believe it or not, before babies can communicate with words, they can notice collections of objects both visually and auditorily. Even more outstanding is that babies (usually around 6 months of age) can recognize that *eight* items is *more* than *four* items, even without knowing the names for the numbers (Libertus, Feigenson, & Halberda, 2011)! This sounds crazy, but evidence from various research studies shows that humans are born with an innate ability to recognize the difference between two quantities through visuals or sound. Researchers call this beginning stage numerosity, which is just the idea that humans (and animals, too!) can notice differences among the number of things in a set. Our brains are wired to estimate the size of a group without having to rely on language or symbols. This beginning approximate number system, as experts in the field call it, is the foundation to our development of number sense.

SUBITIZING

Before children can count (1, 2, 3, 4 . . .) they first recognize general quantities, or amounts. For example, look at the quantity of dots in different formations in figure 3.2.

Once children have explored the quantities by just seeing dots, they then need to attach an amount to give the quantity meaning. Take a look at arrangement A in figure 3.2 and think about how you saw the quantity of five dots. Most likely, you just "knew it" or instantly recognized it. This is called subitizing.

There are two types of subitizing: perceptual and conceptual. By looking at the dots in arrangement A and immediately recognizing "five," you used

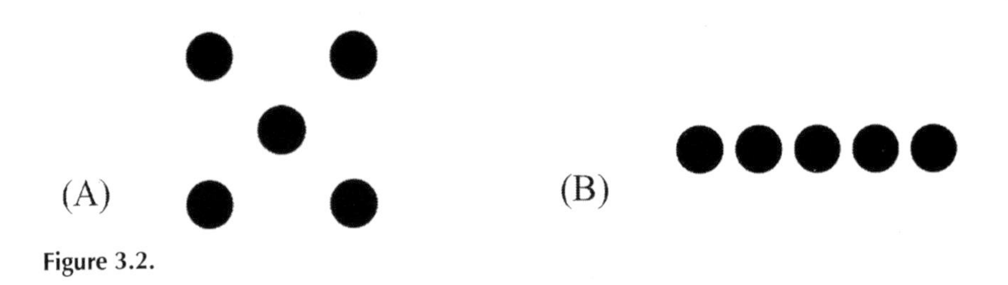

Figure 3.2.

perceptual subitizing. Perceptual subitizing is when you can attach a quantity to a visual and your justification for the quantity is generally, "I just knew it" or "it looks like the pattern on dice, and I am familiar with that."

Now, shift your view to arrangement B in figure 3.2, which also shows five dots. Here, five may not have immediately jumped out at you. You may have had to **partition**, or break apart, the group into parts that you do recognize. For example, you may have immediately seen two dots on the left and three dots on the right and because you have a deep understanding of quantity, you could form that together as five. This is called *conceptual subitizing*. The ability to conceptually subitize does not come until later in the trajectory.

Being able to perceptually subitize prepares children for comparing quantities to make sense of numbers and eventually leads to conceptual subitizing. But, parents beware, spatial arrangements of sets (the way the objects are organized) can make subitizing very difficult. In this learning progression, at this stage, a student may only recognize familiar patterns organized in rectangles, lines, or patterns found in life, such as the arrangement of dots on dice or dominoes.

Take the images in figure 3.3, C and D, for example. Is that easy to immediately see the total number of dots? You most likely had to count at some point. Our brains are not wired to hold more than seven pieces of information at a time (Miller, 1956), and children at this stage are not yet able to count with meaning. Additionally, certain spatial patterns are hard for us to recognize, such as scattered dots or dots arranged in a circle.

We believe at this stage of the learning progression that students should only work to automatically see groups of one to five in familiar patterns. If your child is at this stage, we encourage you to point out common numerical patterns around you. For example, if you are playing a board game with dice, be sure to name the number when you roll and bring attention to the pattern. If you have a deck of cards at home, play a card game and highlight how the quantity of four is arranged on the card.

If you're looking to strengthen your child's ability to recognize quantities, start out with familiar patterns such as arrangements on dice, dominoes, or

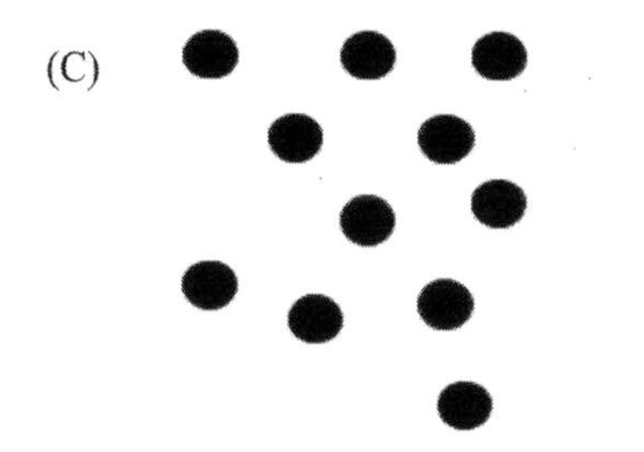

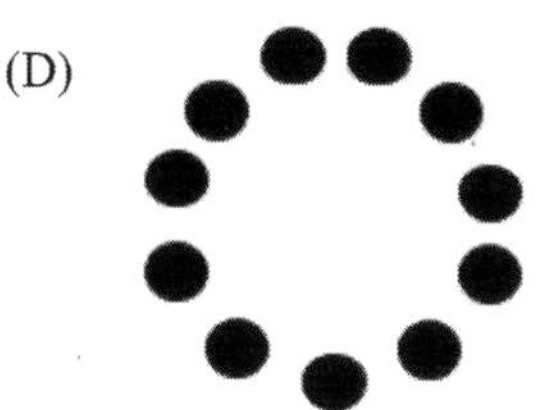

Figure 3.3.

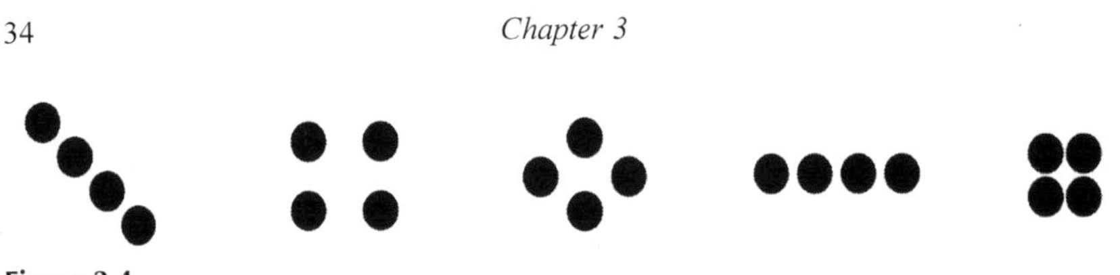

Figure 3.4.

playing cards and stick to seven or fewer objects within the quantity. Represent the patterns on an index card or piece of paper and quickly flash the image so your child has enough time to see the pattern but not enough time to count the dots.

Ask your child, "How many dots did you see?" If your child struggles, start with a smaller quantity (e.g., if you started with four dots, go to three dots). If your child seems to recognize familiar spatial patterns such as dice arrangements, then continue working with quantities fewer than seven, but arranged differently, as shown in figure 3.4. Once your child becomes comfortable with naming quantities, they will then start to compare those quantities to others.

COMPARISON

Now that children have started making meaning with quantities, they will begin to see relationships between these quantities (more than, less than). Notice how we said *quantities* instead of *numbers*. At this stage, children have not yet made a connection between the three components of a number, meaning they do not have a strong enough sense of what the number represents to truly associate meaning to it. Children at this stage are not writing signs for comparison, such as >, <, or =, nor are they able to tell you yet that "four is one more than three." Instead, they are physically touching objects, such as looking at four cubes next to a group of three cubes, and deciding which has more, as shown in figure 3.5. They can justify it by comparing the lengths of the stick of connected cubes.

As children begin to become familiar with quantities it is important to begin adding labels to the numbers. For example, four *eggs* is not more than three *dozen eggs*. Context matters!

At this stage, we encourage parents to say the words *more*, *less*, *fewer*, and *greater*, when appropriate. The kitchen is a great place to do this. For example, if your family is eating pancakes for breakfast, be sure to point out who has *more* pancakes than another, or who has fewer. If you're at the park, have your child tell you if there are more people on the swings or in the sandbox, or if there are fewer people. Again, the idea is to get your child noticing collections of things and seeing relationships between them.

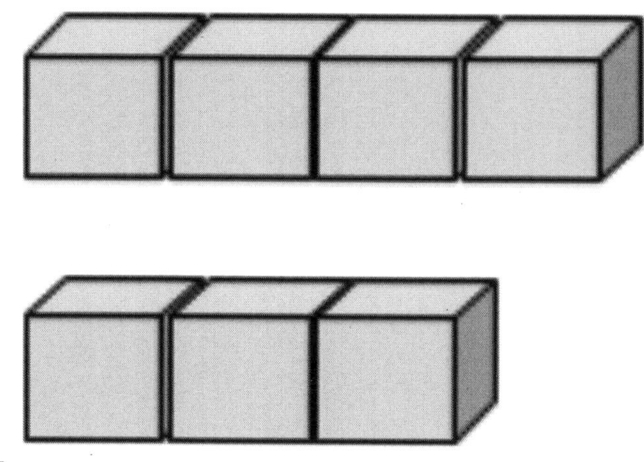

Figure 3.5.

COUNTING

Going back to counting, it is important for you to understand that just because your child can say the counting sequence, e.g., "1, 2, 3, 4 . . . ," does not mean that your child *understands* what it means. When your child is *procedurally counting*, this means they are just memorizing a sequence and regurgitating it. Here, kids learn that counting numbers starts with *1* and that *2* always follows in a forward counting sequence by ones. It is not until the next stage that children start connecting the sequence to understanding.

It is critical that your child practice counting as often as they can, but be certain that you are asking them to count when there is a reason to count! Thinking back to chapter 2, we want to also develop a growth mindset and positive outlook of math. One of the worst things you could do is force your child to count when there is really no reason to do it. So, the next time you go to grab cookies for your child, try instead to ask *them* to get a certain number of cookies (e.g., "You may have *three* cookies. Can you get *three*?"); this way, they are practicing counting but with a reason.

Be sure to give your child different experiences with numbers. Some parents focus so much on verbal counting (counting out loud) that their child becomes great at counting out loud but lacks the deeper understanding and connections to the quantities and symbols, making it more challenging for children to develop number sense.

Since children do not connect meaning to counting until they've shown one-to-one correspondence, be sure to give your child various experiences with all three aspects of number (number words, quantities, and symbols).

ONE-TO-ONE CORRESPONDENCE

At this stage, children learn that to accurately compare quantities, they need to assign one name to one object. This is called one-to-one correspondence. Look at figure 3.6. Notice how a child has to know the rote counting sequence and now attach it to objects. Can you see how that is slightly more advanced than the comparing stage where they are just *looking* for visual differences or the counting stage where they are just naming numbers in order? When pointing to the first object, they say "one." When pointing to the second object in the set, they say "two." Take a moment to think about how hard that task is for emergent counters. They are saying "two" but only touching *one* thing. We as adults recognize that when they say "two," it means there are two objects in the set, but at this stage of development kids don't have that meaning attached. They simply understand that they must recite the counting numbers in order as they touch each object.

This is also the stage where kids understand that length does not always help to determine whether a quantity is more than or less than another quantity. In images E and F in figure 3.7, students need to understand that even though the second row of cubes is longer, there are actually the same number of cubes as in the first row. They know this now because they can line up the cubes in a one-to-one relationship and notice that for every one cube

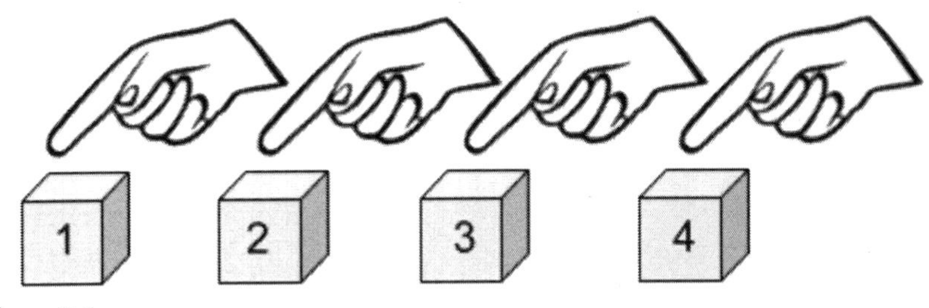

Figure 3.6.

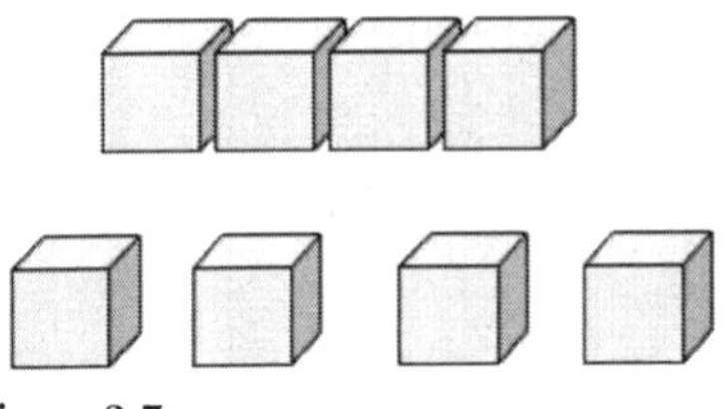

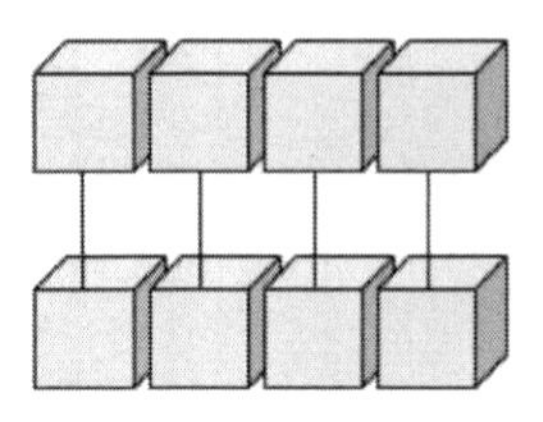

Figure 3.7.

in the top row there is one cube in the bottom row. Examine images E and F to see what one-to-one correspondence looks like through matching objects.

Believe it or not, at this point children still don't have a full understanding of counting. Children need to satisfy one more stage before being able to count with meaning.

CARDINALITY

Children learn how to count (reciting number words with matched number of objects) before they can connect the idea that the last word stated in a counting sequence represents the total number in the set. This is called cardinality. Look closely at figure 3.8.

Imagine a child who is a beginner counter counting to five. You may hear them recite "one, two, three, four, five," while tracking the count on their fingers. That's because they have not yet connected that when they hold up their hand, they have five fingers in total. If after counting the five objects, in this case fingers, a child can answer, "How many do you have in all?" then they have been said to have learned cardinality. If they cannot answer, they will most likely resort back to counting from one. *This is okay!* It just means that they are not yet ready to develop cardinality. Cardinality is an important step in understanding counting, so if your child is working toward it, let them recount as often as they need, but be sure to emphasize when they get to that last number counted that this number represents all the objects in the set.

The idea seems simple to us as adults, but it is actually a very challenging concept for developing math thinkers. Not only do they have to match a number word to an object, but they also have to track how many total numbers are being stated. Many children begin by doing a "triple count." We call it that because they have to start the count from the beginning three times before they trust that the number is correct. Have you ever seen your child do this? Here is what it looks like:

Adult: (*Putting out four objects in a row*) How many are there?
Child: (*Counting and touching objects*) 1, 2, 3, 4.

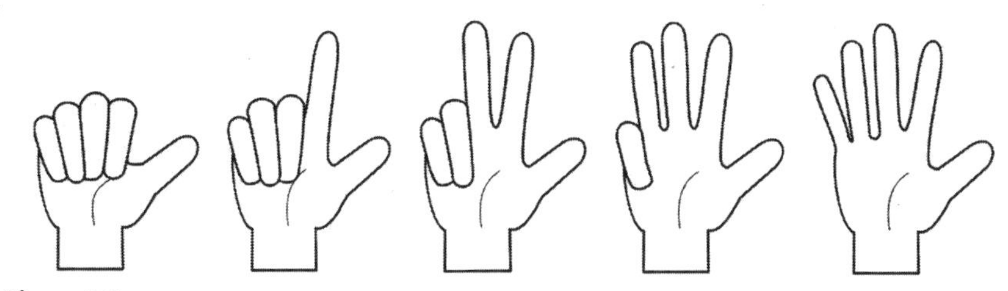

Figure 3.8.

Adult: So . . . how many are there?
Child: (*Counting and touching objects*) 1, 2, 3, 4.
Adult: So . . . how many are there?
Child: (*Counting and touching objects*) 1, 2, 3, 4.

It's easy to get frustrated at this point! As an adult, it seems so obvious that there are *four* objects. But we have to remember that our knowledge base is already built. We subconsciously subitize (just see it), yet children in this stage are not associating the quantity to the number name and haven't quite understood that the last number they said represented the quantity of the set.

Learning to understand cardinality is not an easy task for children as they are developing number sense. Research shows that most children make this cognitive connection by age 4.5 (Fosnot & Dolk, 2001). Remember, *most* does not mean *all*. Don't worry if your child isn't yet able to count a set of objects and retain that the last number they have said represents the total number in the set.

We can't push kids to be where *most* kids are at a certain stage; each and every learner is different, and we have to teach kids where they are, not where we want them to be. One example of this is when parents push kids to recognize that they have five fingers on one hand. Until a child is ready to connect the quantity of five fingers to the number word "five" and the symbol "5," then asking them to move faster by just knowing that there are five fingers on one hand is not appropriate.

HIERARCHICAL INCLUSION

Now that children have cardinality, they are ready to show understanding that **cardinal numbers**, or numbers that represent quantities, represent one more than the previous number. In other words, five objects is one more object than four objects and four objects can never be five objects. Look at image G in figure 3.9 to see a commonly used diagram that illustrates how numbers

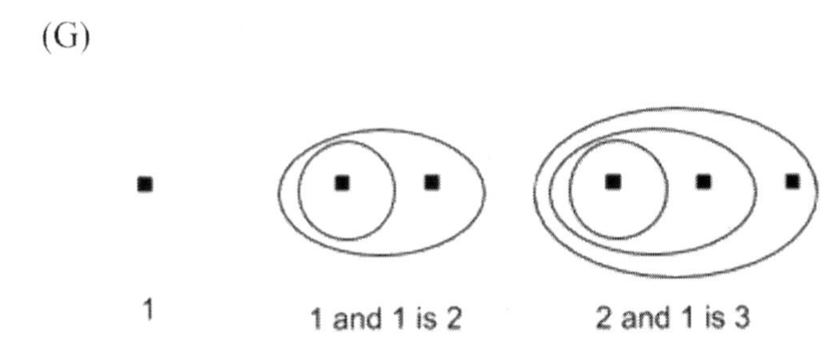

Figure 3.9.

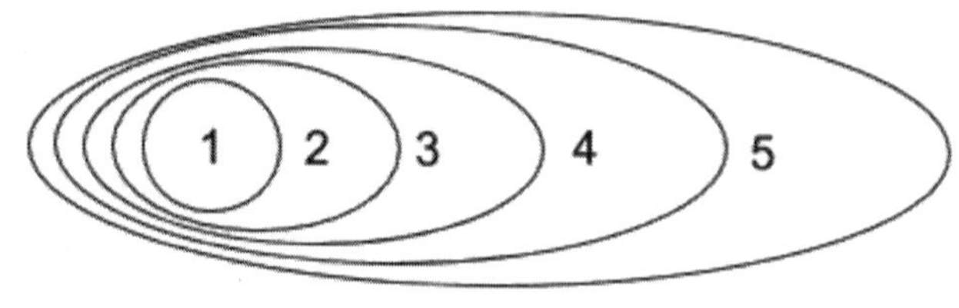

Figure 3.10.

are *nested* within each other. Alternatively, we could represent the same idea abstractly, using just symbols. Look at image H in figure 3.10.

As you can see from the diagrams, if you have two objects, then you have the first one and one more. If you have three objects, then you have the first one, the second one, and the third one to make three. This concept is challenging to understand, so let us show you what hierarchical inclusion *isn't* and then what it *is*, visually. Imagine in the image in figure 3.11 that the speech bubbles on the right are the adult and the speech bubbles on the left are the child.

What hierarchical inclusion *isn't* is shown in figure 3.11. You can see that the adult is asking the child for *three* objects, but the child is stuck in the cardinality stage. The child hears "three" and goes back to the rote counting sequence and identifies the third object in the counting sequence. To the child, that object *is* three! To an adult, it's the third object in the set, but not the quantity of three, as it does not include the first and second smiley face.

Now let us look at what hierarchical inclusion *is* (see figure 3.12). You can see that this child is able to recognize that while the third smiley face counted represents the number three, that in order to satisfy the request, the child must include the first and second smiley faces to make a set of three. *This* is hierarchical inclusion.

It may be helpful to relate the idea of hierarchical inclusion to a very popular toy, a matryoshka doll. Alternatively named "nesting dolls" because the dolls are *nested* within the next largest, matryoshka dolls hold the same idea as hierarchical inclusion. If you don't know the toys by name, take a look at figure 3.13 for a visual of the dolls.

These toys are a great way to introduce the idea of hierarchical inclusion to your child who is ready for this stage. Let your child play, explore, and notice that the larger the doll, the more dolls it can hold inside it and the smaller the doll, the fewer dolls it can hold inside it. After showing understanding of *more* and *less*, children will begin to start showing evidence of flexibility with numbers, leading to the final stage in the trajectory.

Figure 3.11.

Figure 3.12.

Figure 3.13.

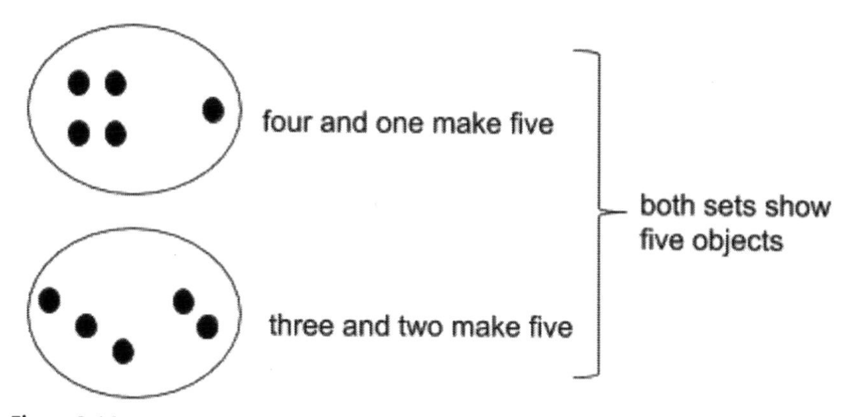

Figure 3.14.

NUMBER CONSERVATION

Once children begin to comprehend that quantities are made from other groupings of quantities, then they can start making sense of the fact that five objects can be represented in many different ways. The number 5 can be represented by a group of 1 and 4, 2 and 3, or 5 and 0. Children who recognize this are then ready to connect that a number can be represented in multiple ways. Figure 3.14 shows what number conservation looks like.

Notice how numbers are more than just symbols; it is critical that children develop meaning behind the symbols in order to become flexible with them. Helping children understand that numbers are made up of parts will lead your child toward operating with numbers using multiple strategies, allowing them flexibility with numbers.

HOW TO USE A NUMBER SENSE TRAJECTORY AT HOME

So, now that you have a deeper understanding of a number sense trajectory, what could you do at home to help your child move past *procedural counting* and move toward *conceptual counting*? Below is an example of a conversation you could have with your child during bath time.

Parent: Let's count the rubber duckies that are in the tub! *(As child counts, point to each duckie from left to right to model one-to-one correspondence.)*

Child: 1, 2, 3, 4, 5.

Parent: *(point to each duckie from left to right as you repeat)* 1, 2, 3, 4, 5. The last number you said was 5, so that must be the number of duckies that you counted! Good job! Let's count to 5 on our hands. *(Hold up one hand and model counting to 5 on one hand, lifting one finger up for each count.)* Can you count to 5 on your hand?

Child: 1, 2, 3, 4, 5.

Parent: Great job! Guess what? One more duckie came to play! *(Put one more duckie in the tub.)* How many duckies are there now?

In this conversation, note the following:

- *If child* **can** *say "6!" then say:* Wow! You must know that six is one more than five (*to reinforce hierarchical inclusion*). How else can you prove that there are six? *(Pause. Can your child say, "count them"? If not, say, "we can count them!")* 1, 2, 3, 4, 5, 6.
- *If child* **cannot** *say "6!" then say:* Let's count together! *(Point to each duckie as you count with your child.)* 1, 2, 3, 4, 5, 6. The last number you said was six, so there must be six duckies (*to reinforce cardinality*)! We can also check by starting from the beginning and counting all the duckies to be sure (*to reinforce counting sequence*).

Continue on with this activity for as long as it takes for your child to start to recognize that every time one more duckie comes to play, the quantity is represented by the next *cardinal* number. Remember! Anything you do forward can be done backward, so be sure to start the count from six and count backward as well.

Math is everywhere, so training your child to recognize math in places outside of school is a great way for you to help your child develop number sense at home. For example, have your child count up and down stairs as they walk or play "I Spy" where you spy a number or collection of things somewhere and have your child try to identify what you are thinking of.

For even more at home resources, see chapter 9.

Chapter 4

Tell Me in Layman's Terms

Speaking the Educational Jargon

> "Making mathematics accessible to the educated layman, while keeping high scientific standards, has always been considered a treacherous navigation between the Scylla of professional contempt and the Charybdis of public misunderstanding."
>
> —Gian-Carlo Rota

We have heard time and time again that parents today feel like they are speaking an entirely different language with their children when it comes to discussing math. The most common complaint we hear is that not only do the actual math calculations appear different from the way most parents learned them in the past but also the way in which kids *talk* about math feels like it is from another planet. Additionally, teachers are so involved in education that often during parent-teacher conferences or communicating with families, they use the educational jargon that has become second nature to them. Unfortunately, parents who are not in educational professions find the language obscure and challenging to understand.

We recognize that these issues create barriers between parents and children in being able to discuss what the children have learned at school. This chapter is meant to help parents have a foundational understanding of the language in which teachers speak. Our goal in laying down the primary vocabulary and tools for parents is to assist in communication between parents and school and parents and children.

As we explore common tools your child might be using in their math classrooms, we will also break down the educational vernacular. We believe it's important that parents know about the tools their children are using in the classroom so they can use them at home to help their children communicate mathematically. Equally as critical is understanding the language used today.

It's important to us that parents feel as though they are speaking the same language as their children.

This chapter focuses on commonplace tools used throughout elementary math classes. We have broken this section down into two parts: manipulatives and representational models. When you hear the word *tools*, you most likely are envisioning instruments such as hammers, screwdrivers, and other implements used to carry out a particular function. When math educators use the word *tools*, we are kind of referring to the same thing—devices that help us solve a problem.

In the math world, physical tools are referred to as **manipulatives**. However, in math, tools don't always have to be physical. Some tools we use are representational, meaning they *represent* physical tools. Think back to chapter 1 where we discussed the Concrete-Representational-Abstract (CRA) approach. Students should first use manipulatives and hands-on, concrete objects to learn a math concept and then, later, represent the physical tool with a diagram or picture. Let's take a closer look at these two types of tools used in today's classrooms and dive deeper into how they are used.

MANIPULATIVES

The word *manipulative* may sound fancy, but it's an important term used in math education today. Let's look at why we call certain tools in mathematics manipulatives. Referring back to chapter 1, you can recall that the first developmental step in learning mathematics is through **concrete**, or hands-on, methods. Imagine you have a newborn baby. Can that baby draw pictures yet or represent their thinking? No, and you wouldn't expect an infant to! It wouldn't be *developmentally appropriate*.

What you would expect is that the newborn will *touch* and *feel* and *explore* with any object. When humans first learn, they do so through **kinesthetic** means—basically, they first learn best by touching and feeling! So, when it comes to learning mathematics, studies have shown that children (and adults) learn best by *manipulating* hands-on objects.

Manipulatives allow children to touch, feel, and visualize what they can't yet create on their own. Manipulatives also allow children to receive immediate feedback about whether an idea makes sense. A child can *move* hands-on objects to *explore* and *investigate*. For example, a topic that is often challenging for children is finding the surface area of an object (the area of the outer part of an object). While a child can certainly see this through a drawing of a net, it is much easier to visualize by first taking an actual box and breaking it apart to identify the area of each face, or side, of the solid, or 3-D figure (see figure 4.1).

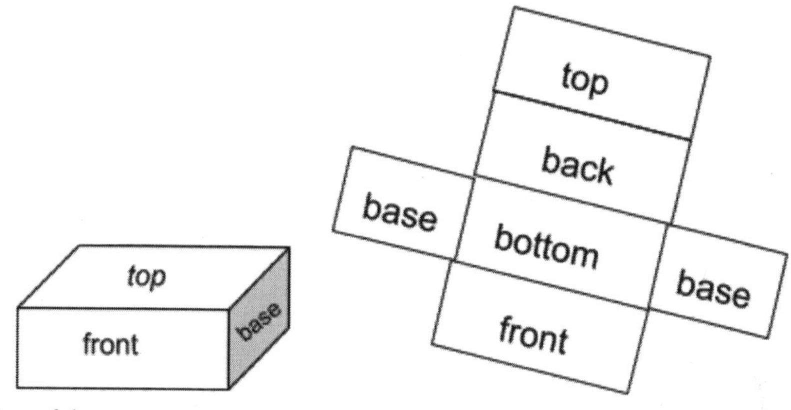

Figure 4.1.

In this chapter, we will introduce some common manipulatives used at the elementary school level so you can gain more familiarity with the tools your child uses daily. Even though these are educational manipulatives, we highly recommend you purchase sets for your home. Kids should be using these tools at home to support their learning in school. Have your child *show* you what they learned in math class, rather than asking that age-old question, "What did you learn today?" and receiving an "I don't know" or "nothing" as a response.

We have tagged each tool with its beginning grade level, but be sure to read closely as some tools can be used well beyond their rudimentary setting. A simple internet search will curate many places from which you can order the materials. Also, we have listed a few ways in which you can use these tools at home to support your child's math development, but know there are *many other ways* to use these tools. Again, use the internet to find games and activities you can play at home in addition to the ones we mention.

In the following chapters, we will provide specific examples of how students will use many of these tools to add, subtract, multiply, and divide whole numbers and fractions.

Unifix Cubes (Preschool +)

Unifix Cubes are great manipulatives that have such a long-lasting use that investing in the tool for your home is worth it (see figure 4.2). Children from the preschool age through middle school can find use for these blocks. Below, we show some examples of how to use these tools in the elementary grades. *Please note: You can use other types of cubes for these activities, but we advise Unifix Cubes because they can link together easily.* Keep in mind that these tools have many more uses, and some can even be used in literacy and other subjects! We

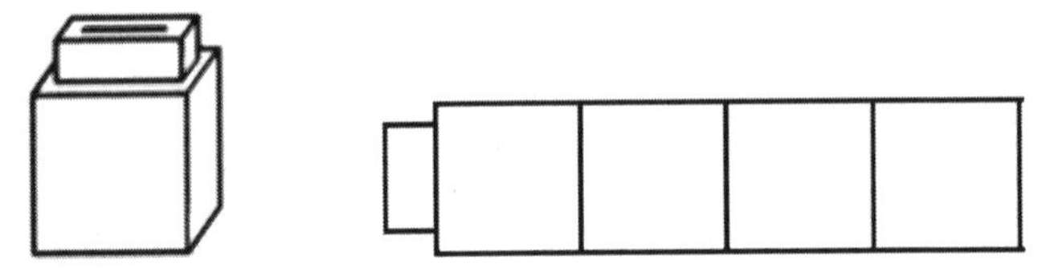

Figure 4.2.

aim to provide you with a few examples to get started, but we encourage you to search the internet or talk to your child's teacher for more ideas.

Preschool

Children ages 0 to 4 should use these cubes for playtime. The cubes vary in color, which is helpful for children who are just learning their colors. Use the varying colors as a place to start mathematically—have your youngster sort the cubes by color. Mathematicians organize through sorting and categorizing, so this is a good first activity for young children.

Unifix Cubes also help children develop fine-motor skills (the ability to use small movements and muscles) when children link the cubes together. Mathematically, children ages 2–4 may use them as objects to count. By putting the cubes together, children are subtly exposed to the idea of addition as joining, and inversely, by taking the cubes apart, children are playing with the idea of subtraction as take away. Children at this age may also recognize patterns and put alternating colors together.

Kindergarten

Children around age 5 should spend a lot of time at home counting cubes. Use the cubes to start noticing how your child responds to different spatial patterns. For example, lay six cubes in a line and ask, "How many cubes are there?" Then, lay six cubes in two groups of three as you'd see on dice and ask, "How many cubes are there?" Now try scattering the cubes so they are not in any recognizable order. This may be much more challenging for your child and if not, then you may see your child exhibiting **one-to-one correspondence** as they touch one cube and say "one" and then drag another cube to be in the same row and say "two" and so on. Refer back to chapter 3 to review one-to-one correspondence.

While your kindergartner practices counting, you can also have them use the cubes to build and represent the quantity of five (and so on to 10, as appropriate). For example, give your child five cubes. Ask your child to make two groups with the cubes by saying, "Can you put *some* cubes here [point to one place] and *some* cubes here [point to a different place]?" For example, a child may place three cubes in one area and two cubes in another. Keep practicing other ways to represent the quantity of five (e.g., 4 and 1; 2 and 3; 5 and 0). This helps build a foundational understanding of the concept of part-part-whole, which will be explored further when discussing **number bonds** in *representational tools*.

Another way to use the cubes to support developing understanding of the number system is to have your child practice connecting 10 cubes together. At the kindergarten stage, it is critical that children begin to see that 10 separate ones (in this case, cubes) can be grouped together to represent one group of 10. This concept is called **unitizing**. When students unitize, they recognize that many small units can be regrouped to represent one larger unit of equal value (e.g., 25 pennies is the same as one quarter). This notion will become very important later in your child's development of mathematics, as it provides foundational understanding of place value.

So, for now, have your child connect 10 cubes and say "ten" and then represent the quantity of 11 with a connected stick of 10 and one separate cube (so they can see 11 is really a 10 and 1). Repeat this exercise through the number 20, where your child will then see two groups of 10 is 20.

In addition to counting activities, Unifix Cubes are great for beginning understanding of measurement. Play "more, the same, or less" with your child by connecting a certain number of cubes together (we advise sticking to one color) and asking your child to compare it to another group of connected cubes. Be sure when your child is comparing the lengths of the connected cubes that both groups of cubes are lined up edge to edge so your child can use the visual to determine whether one is more, less, or the same. If your child has *one-to-one correspondence*, then feel free to let them be unaligned; this will encourage your child to line up the cubes themselves.

Grades 1–2

Emerging first graders are still often mathematically similar to kindergartners, so don't be surprised if by first grade your child is still using Unifix Cubes to practice counting or identifying the sums to 10 (e.g., 5 + 5; 4 + 6; 7 + 3 . . .). *Remember, this is okay!* Once your first grader has begun to show an understanding of addition as the joining of quantities (primarily by discovering the sums to 10), a new challenge to present is to use Unifix Cubes to help them make sense of their **doubles facts**.

As you may be able to infer, doubles facts refer to addition expressions in which the two addends, or parts, are the same number (e.g., 4 + 4; 9 + 9; 7 + 7). In early elementary school, teachers use terms such as *doubles facts* instead of *doubling* because at this stage the students are thinking additively, not multiplicatively. It also helps students create a visual anchor when they hear "doubles plus one facts," as they describe quantities that are one more than doubling. Don't be alarmed by these terms, but rather, just ask your child to describe them. More often than not, you'll find the terms used today are just variations of terms we used to use!

Using the cubes, ask your child to show you the sum, or total, of a doubles fact. Encourage them to create two stacks of the cubes. This will help them draw connections to repeated addition. Have your second grader begin to look at repeated addition by stacking several cubes together and creating copies of those stacks. See figure 4.3 for a visual example.

Ask your child to describe what their stacks of cubes represent. In the case of figure 4.3, a student might say, "I built one stack with four cubes and another with four cubes. Together, that's eight cubes." Encourage your child to write down the doubles fact, or the equation, they are representing with their cubes (e.g., $4 + 4 = 8$). As you continue this activity with your child, be sure to connect their representations to skip counting. Figure 4.4 shows how to connect to skip counting.

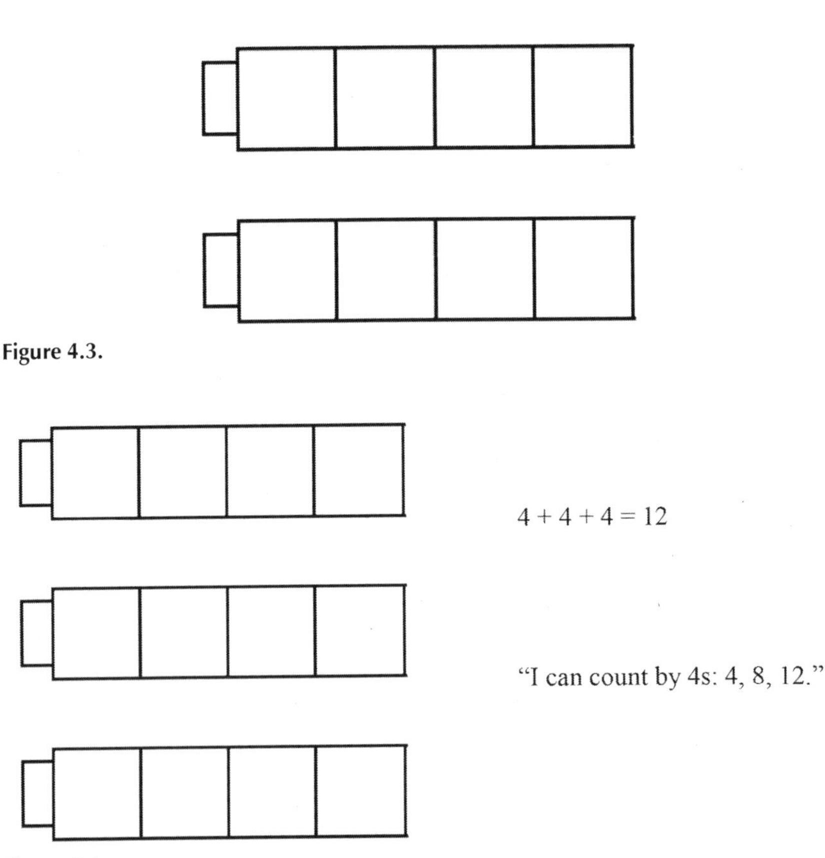

Figure 4.3.

Figure 4.4.

Asking your child to skip count helps them continue to see groups of objects, rather than counting by ones. Counting by ones is a fine strategy, but it isn't always efficient. If your child still leans toward counting a group of objects by ones (e.g., "1, 2, 3, 4 . . . ") then start with adding groups of two objects to help them begin to see that we can "skip" the count of every other number and still say the total that is there. Here's a fun activity (shown in three parts in figure 4.5) to try with your child who cannot count by twos yet:

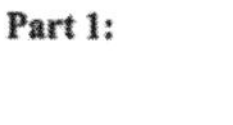

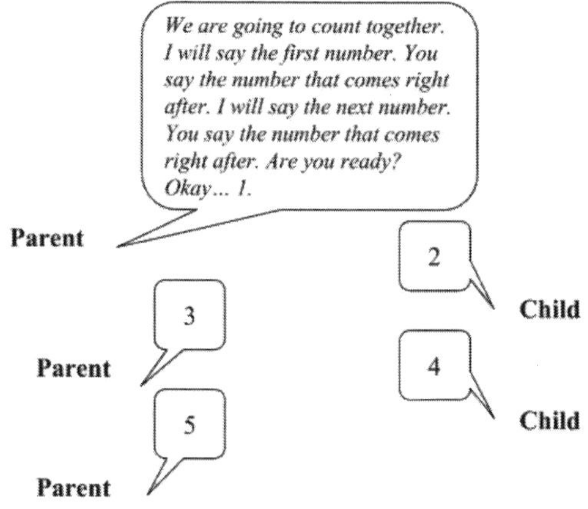

Continue until your child says 20.

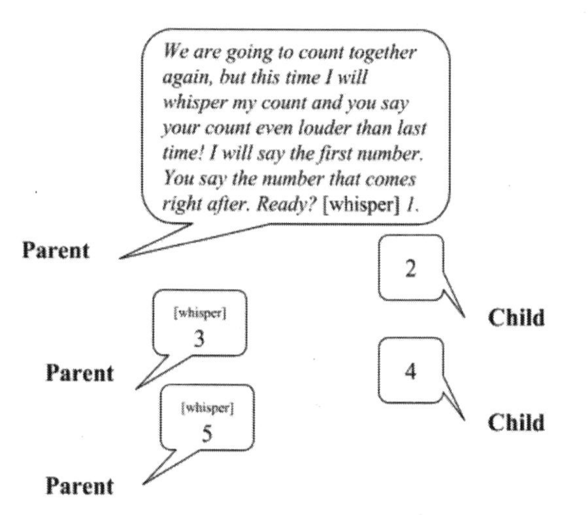

Continue until your child says 20.

Figure 4.5a.

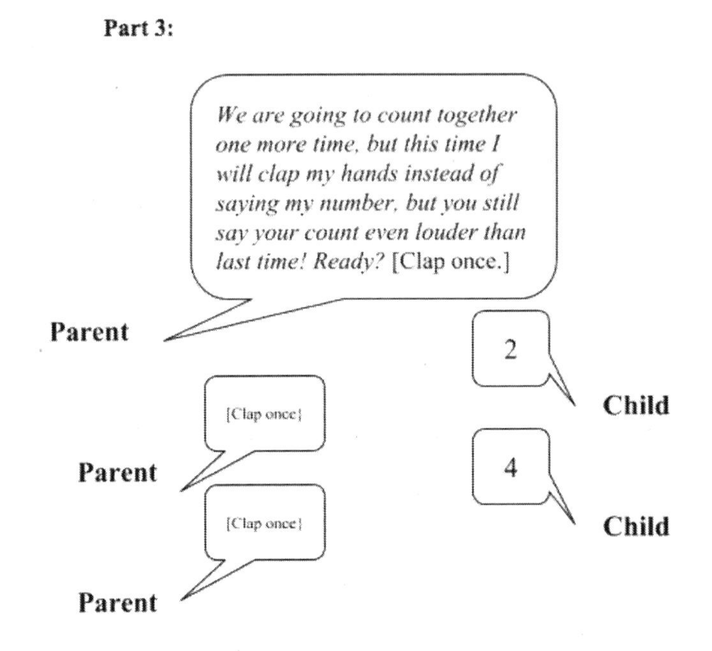

Figure 4.5b.

At this point, your child has counted by twos. Now, take the Unifix Cubes and organize them in groups of two. Ask your child to count by twos again, this time pushing two cubes aside as they count. This will help your child connect the quantity to the number word they are saying. Once your child has practiced this enough to be able to count by twos, place out three groups of two cubes, as in figure 4.6, and ask them to count by twos to find the total number of cubes. Then, have them write the matching repeated addition sentence.

Skip counting by 2s, 3s, 4s, 5s, 6s, 7s, 8s, 9s, and 10s will set your child up for ultimate success in third grade as they begin to develop an understanding of multiplication.

Two other skills second graders should be developing are estimation and nonstandard measurement. An estimation is a close approximation, and nonstandard measurement are units that aren't typically used to measure things (e.g., **standard measurements**: inches, feet, meters; **nonstandard measurements**: paperclips, pencils, a thumb). One way your child could integrate both concepts is by using Unifix Cubes to measure objects around your home. Ask your child to estimate the length of the kitchen floor (in Unifix Cube units) and then ask them to prove it by connecting cubes and laying them on the floor to see how many are needed.

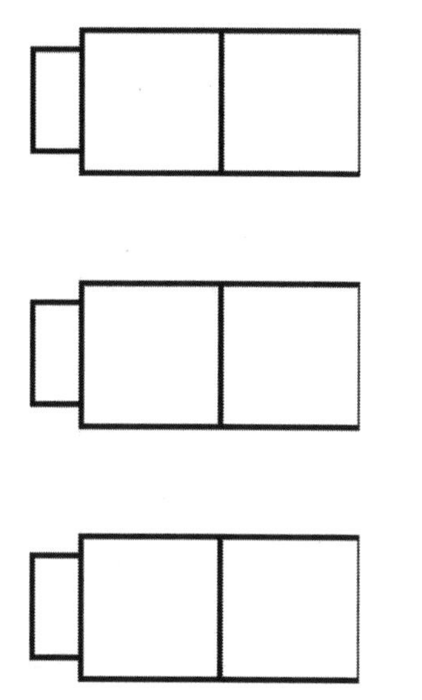

Figure 4.6.

Grades 3–5

Once children leave second grade, a majority of their math time is focused on drawing connections from addition and subtraction to multiplication and division. Unifix Cubes can be used similarly to the way second graders use them, except rather than focusing on repeated addition, children can begin to draw the connection to multiplication by visually seeing groups organized in arrays. Start by asking your child to count by twos and make three groups of two cubes (see figure 4.7).

Continue having your child build various groups of cubes and describe what they see using words and equations. Repeated practice of this activity will help your child begin to internalize the multiplication facts. As they learn their multiplication facts, it is helpful to teach them that division is related. Try the next activity to help your child begin to make the connection between division and multiplication. Hand your child a stick of six cubes, and say the sentence shown in figure 4.8.

Naturally, children will tend to break apart the cubes and share them among three groups (either by ones, or if they see the connection, by breaking the cubes into three groups of two). This is where parents should help their children draw the connection that while they see three groups of two cubes, they started with six cubes and shared them among three groups, resulting in two cubes in each group. Help them make the connection by writing the equations $3 \times 2 = 6$ and $6 \div 3 = 2$.

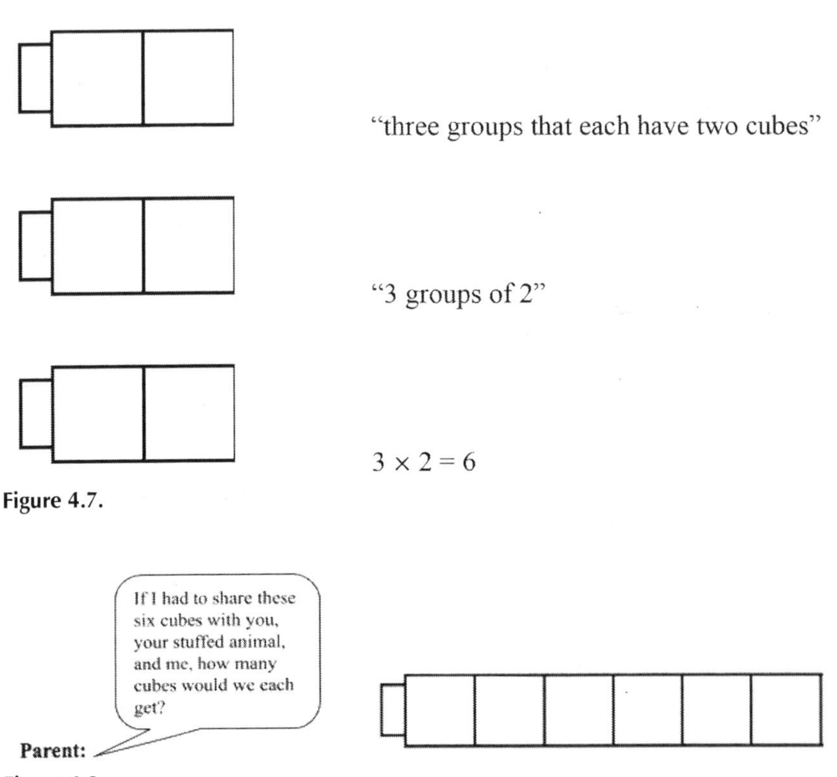

Figure 4.7.

Figure 4.8.

Also, be sure you when you work with your child that you aren't always posing the same type of questions. Notice how the next activity still targets division, but requires a different perspective (see figure 4.9).

Continue practicing so children are getting lots of exposure to related facts. Ultimately, you will want to encourage children to change their organization from groups, to arrays, or rectangular arrangements of blocks. See figure 4.10 for an example.

As children begin to develop a more solid understanding of multiplication and division through seeing rectangular arrays, you can encourage your child to use the Unifix Cubes to start thinking about area and perimeter and drawing connections between all four operations. For example, have them build an array with six columns and five rows, then ask them to find the area (or the amount of the two-dimensional space of the surface). Six columns and five rows results in 30 square units.

When children reach fifth grade, another topic they need to know is how to concretely find the volume, or amount of space, of rectangular prisms (three-dimensional figures that look like a box). Rather than learning a formula in a rote manner, it is important that children make sense of the task. Have your child build a

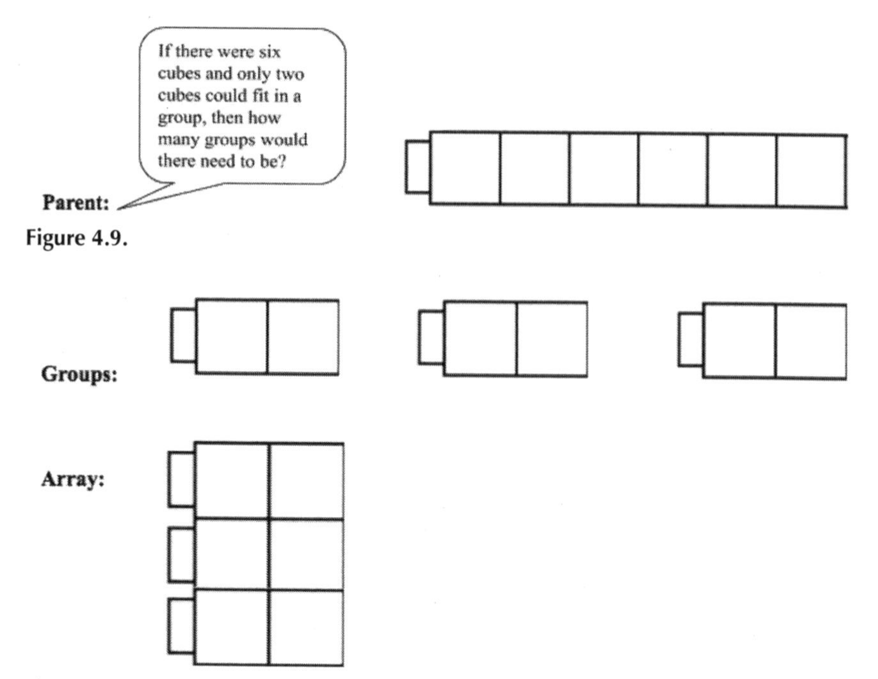

Figure 4.9.

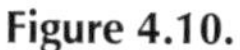

Figure 4.10.

rectangular prism and figure out how many cubes it took to build it. Discuss and draw connections to finding the volume of a figure.

Beyond Elementary School

Use these cubes to support ratio understanding and proportional reasoning. For example, state a rule, such as, "for every two red cubes, there need to be four blue cubes." Then ask your child to determine how many blue cubes there would be if you put out eight red cubes.

> **TIP:** Unifix Cubes are plastic, which means you can write with dry erase markers and they will wipe off! So, be sure to get a dry erase marker for your child and let them write on the cubes! Use the cubes to practice spelling, literacy, and phonics activities, too!

Counters (Preschool +)

Counters are a fun manipulative that are cost effective and have many uses. Counters come in two forms: either dual-colored—meaning that one side is one color and the other side is a different color—or single-colored. If you purchase counters, we advise you to get the dual-colored counters, as it allows the tool more longevity. Usually dual-colored counters are red and yellow, but you can also get them in other colors. The ones your child will be using in school are most likely red and yellow (see figure 4.11).

We suggest using counters very much like how you would use Unifix Cubes, but be aware that the counters cannot link with each other like the cubes can, which limits their use in some ways. While they cannot be linked, they are cheaper to purchase, offering a different advantage.

Because our suggestions for counters are so similar to the usage of Unifix Cubes, we will just show you one example of how you could use the counters in the upper elementary setting.

Counters are a great way to visualize fractions in different ways (remember, you could still do this activity with Unifix Cubes, too). Look at figure 4.12 for one example.

The counters can be placed in a set model and can easily show part-to-whole relationships.

Beyond Elementary School

As children enter sixth grade, they begin to study percentages. Draw connections to their previous fraction understanding to identify what percentage of the set are white counters, for example. Counters are also great beyond

Figure 4.11.

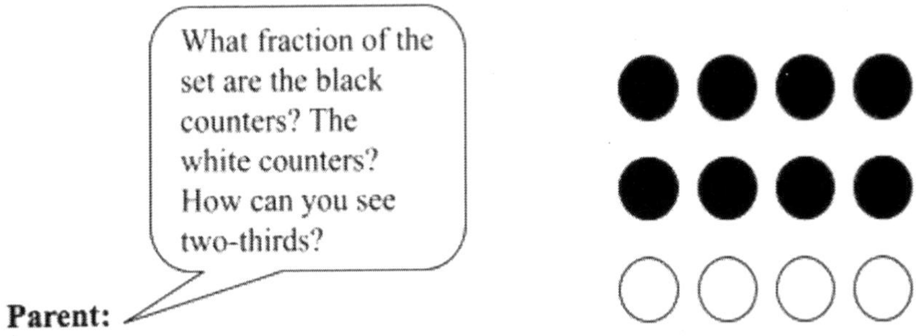

What fraction of the set are the black counters? The white counters? How can you see two-thirds?

Parent:

Figure 4.12.

elementary school for helping children understand operations with positive and negative whole numbers, or integers.

Typically, teachers assign red counters as negative and yellow counters as positive. Feel free to choose your own system at home should your counters be different colors. A positive value counter matched with a negative value counter form a zero pair (e.g., $1 + (-1) = 0$). This important idea is the foundation for understanding how to add and subtract with negative numbers, a topic that frustrates many middle school students who try to memorize a variety of rules.

Pattern Blocks (Grade K +)

Pattern blocks are a type of manipulative that enables students to visually see how shapes can be composed, or made up, of other shapes and also decomposed, or broken apart, into different shapes. A standard set of pattern blocks include six basic shapes (see figure 4.13):

- Yellow regular hexagon
- Red regular trapezoid

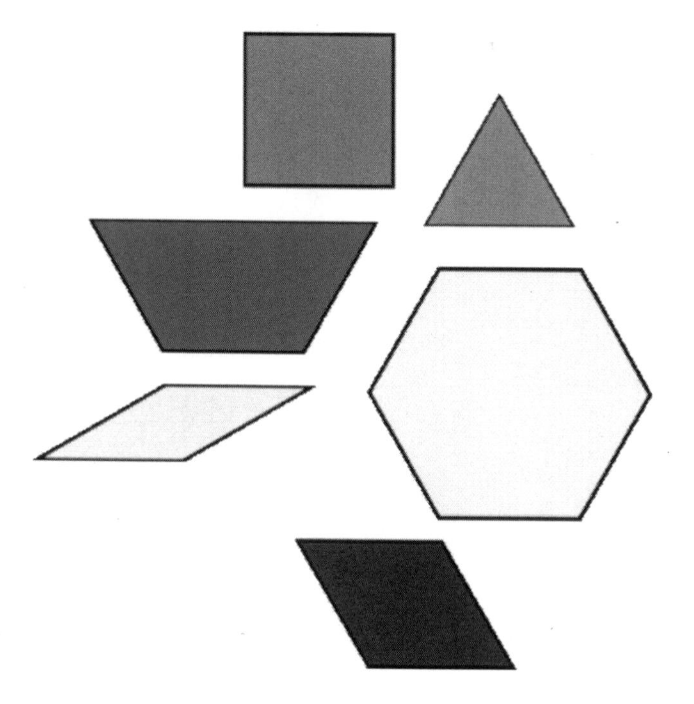

Figure 4.13.

- Green equilateral triangle
- Blue rhombus
- Beige diamond
- Orange square

Pattern blocks are used in many different ways. A few common uses for pattern blocks at the early elementary level are identifying and naming shapes and shapes' defining attributes (what makes a shape a shape), discovering what shapes are composed, or made up of other shapes, and more. At the upper elementary level, pattern blocks are generally used for developing conceptual understanding with fractions, recognizing visual patterns, identifying angles of shapes through geometry, or studying tessellations (tiling of repeated patterns). Let's take a look at a few examples of activities you could try with your child.

Preschool

With your preschooler, use pattern blocks to focus on identifying shapes and various attributes of shapes. It's important to keep in mind that these tools actually represent three-dimensional shapes since they have depth to them. This means that the triangle piece is actually a triangular prism, not a triangle. So, when you focus on two-dimensional shapes, or flat shapes, be sure to just point to one face, or side, of the block, indicating you are talking about the flat part. Figure 4.14 shows an example of how you could use pattern blocks with your preschooler.

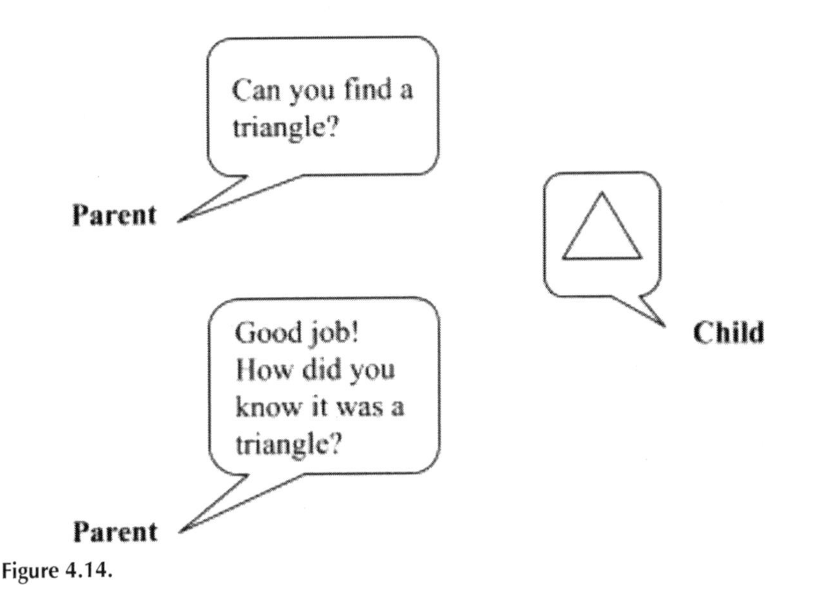

Figure 4.14.

Your child may struggle at this age to communicate precisely how they know it is a triangle, but be sure to add on to their thinking by confirming that the triangle has *three* sides and *three* corners (pointing to these as you say them), reinforcing both the identity of a triangle and its defining attributes.

Grades K–2

Starting in kindergarten, pattern blocks are a great tool to help children see that they can compose and decompose shapes to form other shapes. Ask your child to build a picture and identify shapes within the picture they've built. Emphasize that you want to see different shapes built from other shapes (e.g., a rectangle built from two squares).

Another fun activity for children this age is to build tessellations, or arrangements of polygons in repeated patterns. Look at figure 4.15 for an example. If creating tessellations is a challenge for your child, then do a quick internet search for tessellation pattern block templates and print one out that they can then fill in with their pattern blocks. Eventually, however, you do want them to be able to create a pattern themselves, as this requires higher-order thinking skills.

Grades 3–5

In grades 3–5, a critical area of focus mathematically is on the development of fractional reasoning. Pattern blocks can be a great tool to use to help children visualize fractions. As with all the tools listed in this chapter, a simple internet search will generate hundreds of activities, lessons, and ideas that you can use in your home. In chapter 7, you'll learn more about fraction instruction and see varieties of ways pattern blocks can help you develop a deeper understanding. For now, figure 4.16 shows one example.

Another focal point of fourth-grade math is learning about angles. Pattern blocks are also a nice tool to use, prior to a protractor, to help children make

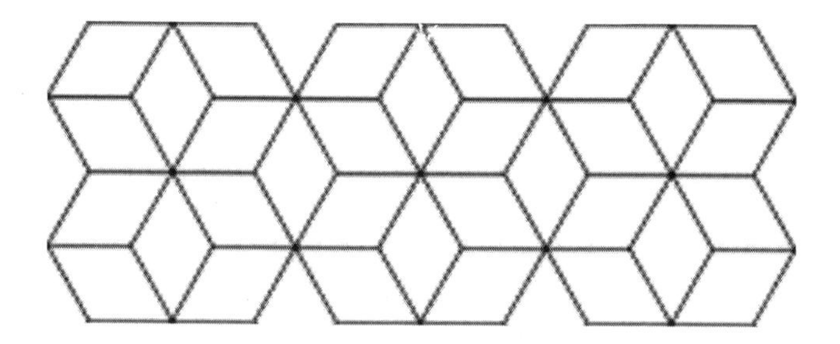

Figure 4.15.

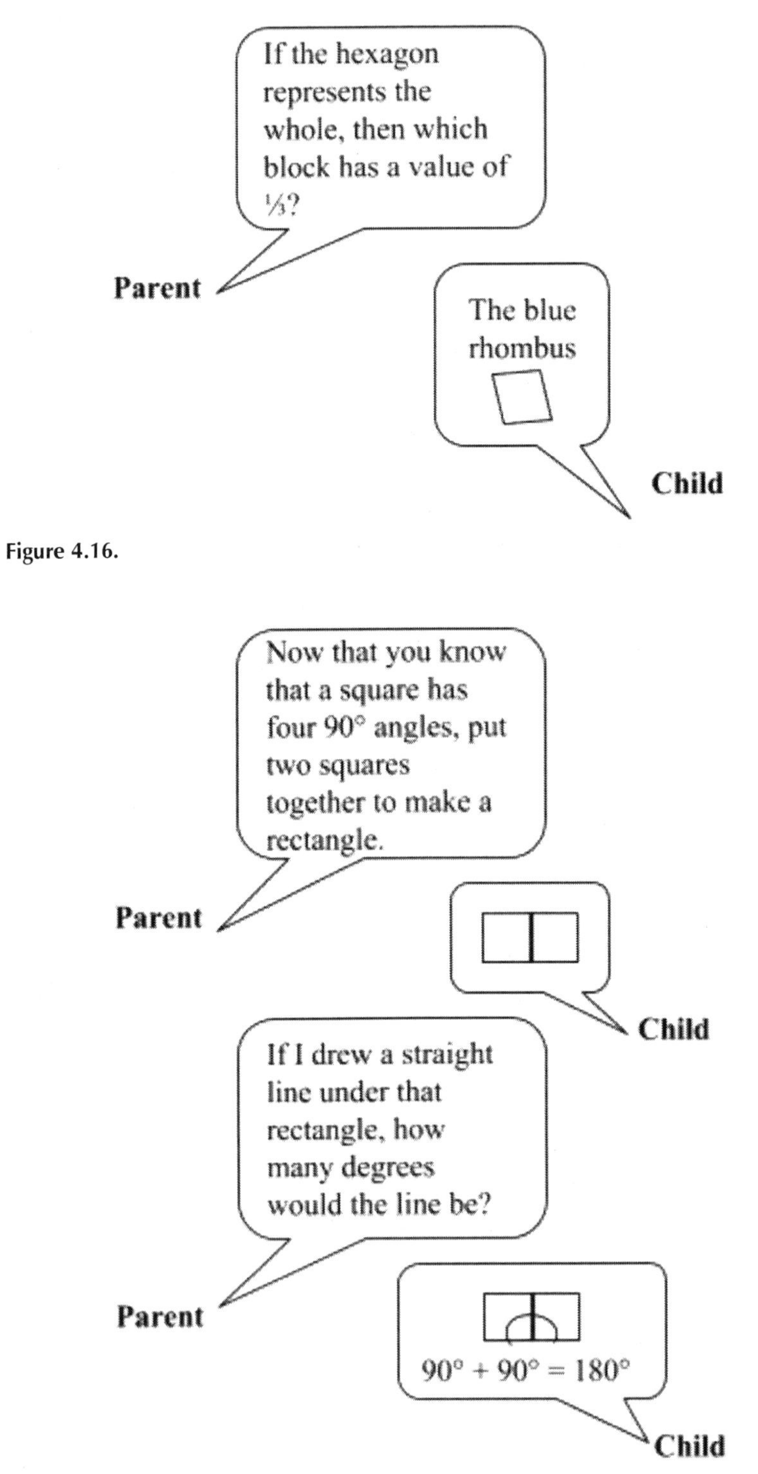

Figure 4.16.

Figure 4.17.

sense of angles. Figure 4.17 shows how you could pose a question using pattern blocks to explore angles. *Note: Your child must already know that a square is defined as having four right 90-degree angles to truly conceptualize the concept of a straight line, as illustrated.*

If your child is following along and can rephrase for you why a straight line drawn under two connected squares is 180 degrees, then we suggest allowing your child to do more exploration. You can do this by saying, "Just as we know that two 90-degree angles make a 180-degree angle, can we do the same sort of thing with triangles instead of squares? What could we do?"

If your child needs more support than this, follow this suggested conversation shown in figure 4.18.

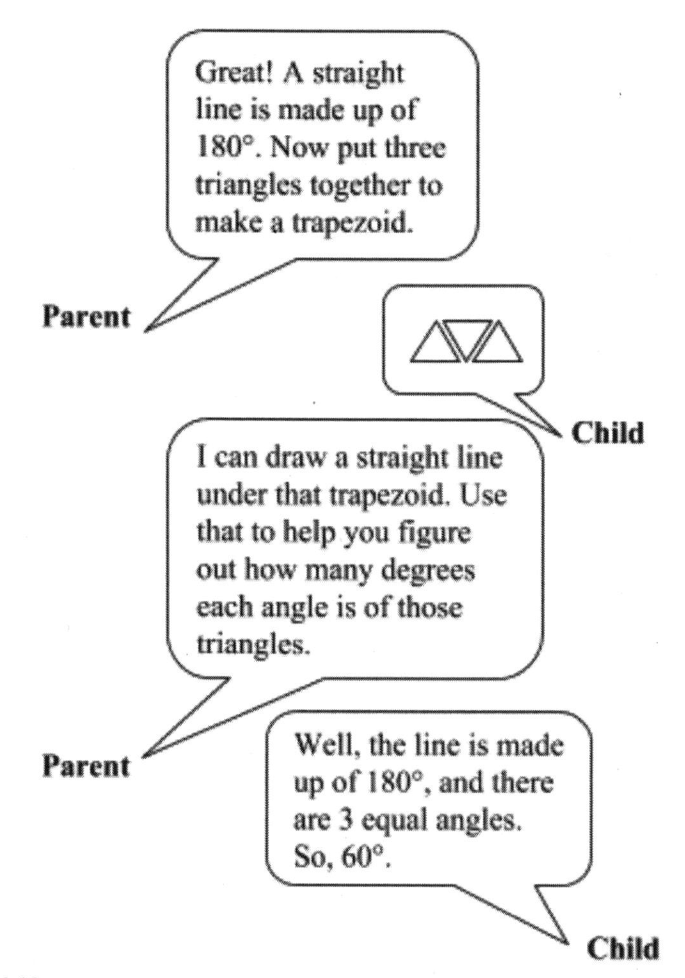

Figure 4.18.

Beyond Elementary School

Children in middle school can continue to use pattern blocks to further explore the relationships they built when they were younger, but this time, using the blocks to generalize patterns and develop rules for all relationships. Also, students can use the blocks to study combinations, transformational geometry, properties of geometric figures, congruence, and more.

Base-Ten Blocks (Grade 1+)

Place-value blocks are also known as Dienes blocks, or multibase arithmetic blocks. They were originally developed by Zoltán Pál Dienes, a mathematician who often used the blocks to do arithmetic in many different base systems, not just base ten. You can find other sets that are meant for other base systems, such as base three or base five, but most commonly used in American classrooms is the base-ten set (see figure 4.19).

Base-ten blocks are tools that help children recognize the value of numbers. These three-dimensional blocks come in wood or plastic and work differently than the Unifix Cubes and counters previously mentioned. When you buy base-ten blocks, you will get a kit with four distinct types of blocks: units, longs (sometimes called rods), flats, and blocks. These four arrangements are meant to highlight place value. In the early grades, students will use the units to represent the ones' place, the longs or rods to represent the tens' place, the flats to represent the hundreds' place, and the blocks to represent the thousands' place.

As children enter fourth or fifth grade and begin to make sense beyond whole numbers, you can reassign the base-ten blocks so that they represent decimals instead. For example, the unit would represent the thousandths' place, the long or rod would represent the hundredths' place, the flat would represent the tenths' place, and the block would represent the ones' place. This can sometimes confuse children more than aid in their understanding of fractions, since they've usually spent three to four years knowing the blocks

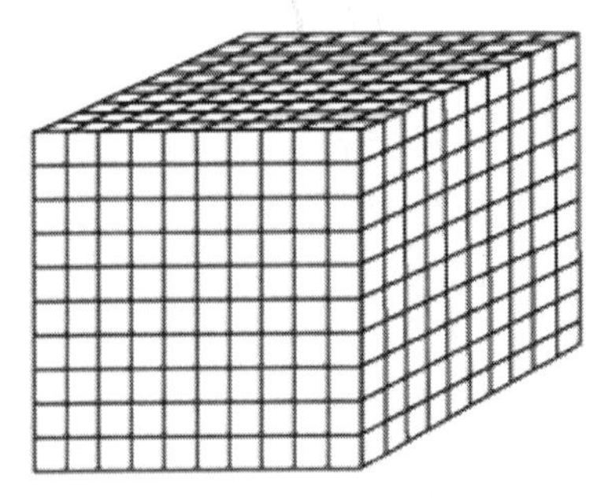
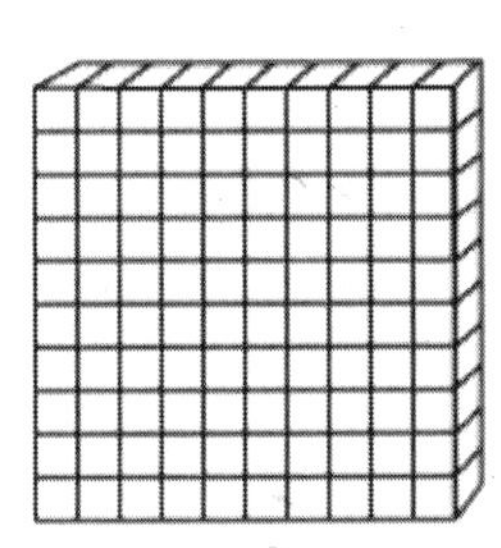
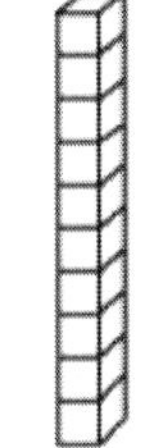

Figure 4.19.

to represent something different, so if you choose to use the blocks in this manner, we suggest spending a lot of time focusing on renaming the blocks.

Base-ten blocks have so many uses that they truly deserve their own chapter. We will show a few different uses for them below, but again, please remember a simple internet search will curate many more ideas for you.

Grades 1–2

Use base-ten blocks to help your child add and subtract whole numbers with a focus on place value. For example, if your child is asked to solve 29 + 34, encourage them to use the blocks to solve. You can structure their learning even more by either ordering or making a place-value chart where they can put the blocks on top of the individually assigned place values and then add or subtract one column at a time. Figure 4.20 shows an example.

Grades 3–5

Use base-ten blocks to help your child multiply and divide whole numbers with a focus on place value. This tool is used best when working with certain factors. For example, if your child is asked to solve 3×52, then using these blocks just as they did for addition works well (it's pretty easy to make three groups of 52 on the place-value chart). However, if your child is working multiplying 14×52, for instance, then using the place-value chart and repeated addition will not be as efficient. They can certainly use this strategy, but as the numbers

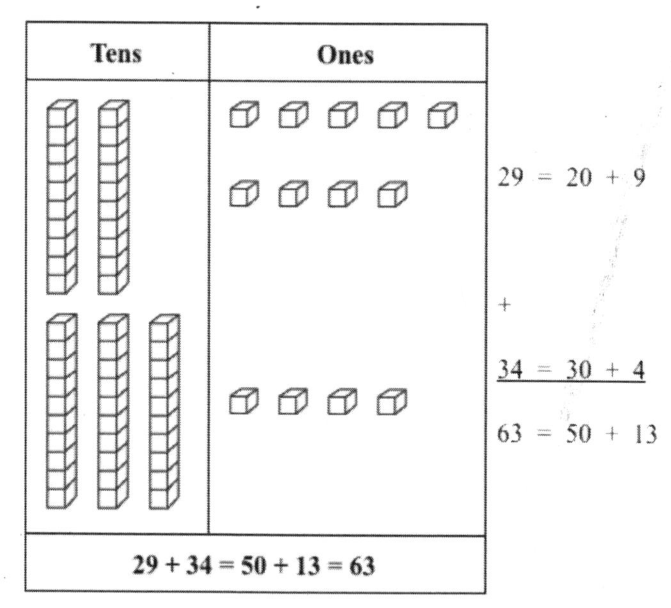

Figure 4.20.

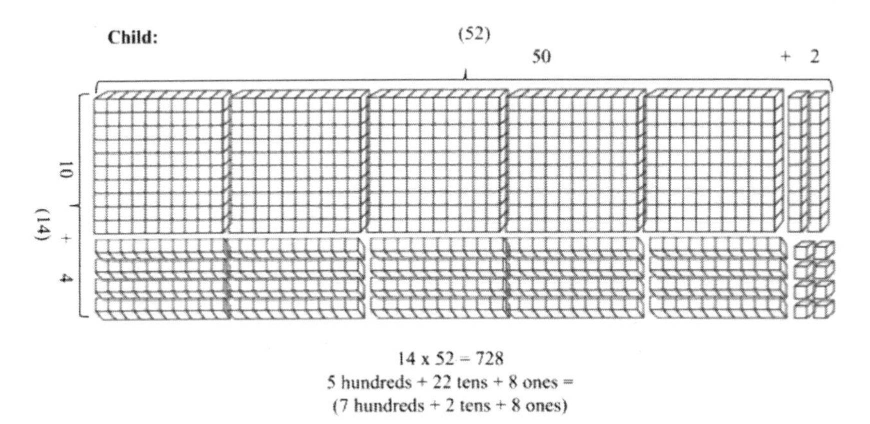

Figure 4.21.

become greater and greater, your child will eventually realize that this tool being used in this manner will take too much time (and require too many cubes!).

So, at that point you will want to transfer over to the area model, a representation in which the length and width of a rectangle represent parts of a multiplication or division problem. We discuss the area model in much greater depth in chapter 6. For now, figure 4.21 shows how you could encourage your child to use the tool to solve 14 × 52.

Using the same philosophy, base-ten blocks can also be used for division. This will be shown in more detail in chapter 6.

Beyond Elementary School

Base-ten blocks continue to be useful in middle school as students develop a conceptual understanding of surface area and volume, operations with decimals and algebraic expressions, and using the proportional relationships embedded within the blocks to solve ratio tasks.

Cuisenaire Rods (Grade K +)

Cuisenaire Rods are sets of blocks in the shape of rectangular prisms, each with its own unique color. The Cuisenaire Rod set comes with 10 main colored rods:

- White (1 centimeter long)
- Red (2 centimeters long)
- Light green (3 centimeters long)
- Magenta (4 centimeters long)
- Yellow (5 centimeters long)

- Dark green (6 centimeters long)
- Black (7 centimeters long)
- Brown (8 centimeters long)
- Blue (9 centimeters long)
- Orange (10 centimeters long)

As with all the other tools mentioned in this chapter, Cuisenaire Rods have a long range of uses that we could not fully cover in our book. We encourage you to use the internet to search for activities and games you could play with your child at home.

Kindergarten

Because kindergartners love to be creative, we encourage you to let your child explore the Cuisenaire Rods by letting them just play with the pieces. Part of their play time will include them looking at the relationships between the colors and lengths. If your child isn't familiar with this tool, then we also advise that after some exploration you ask your child to build a staircase with the pieces (see figure 4.22). This activity helps children recognize that each rod is one unit (in this case, centimeter) longer than another. Once children have successfully identified the staircase, it's a good idea to leave it on the table for reference.

In kindergarten, students are learning how to add and subtract within five, so one activity you could do with your child is to see if you can find all the rods that, when put together, are 5 centimeters long in total (see figure 4.23).

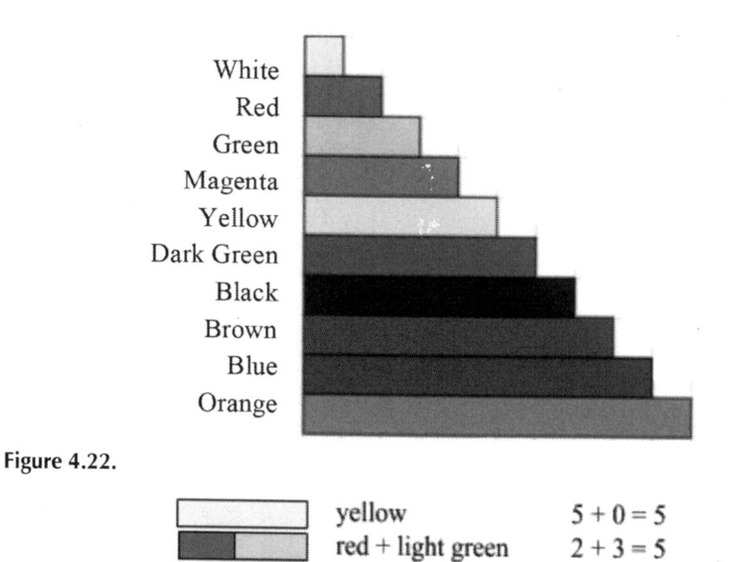

Figure 4.22.

Figure 4.23.

Grades 1–2

Once children start to become familiar with the Cuisenaire Rods and their arrangement, many students begin to show great ease with finding their sums to 5. Next, have them work on their sums to 10. It is essentially the same activity, except now using the other five rods. Figure 4.24 shows some combinations that are 10 centimeters long in total.

As students begin to become familiar with their basic facts and with the rods in general, help them use relational thinking skills by posing some questions similar to the ones shown in figure 4.25.

To make the task more challenging, ask questions that go beyond 10 centimeters in length (see figure 4.26).

orange	$10 + 0 = 10$
magenta + dark green	$4 + 6 = 10$
white + blue	$1 + 9 = 10$

Figure 4.24.

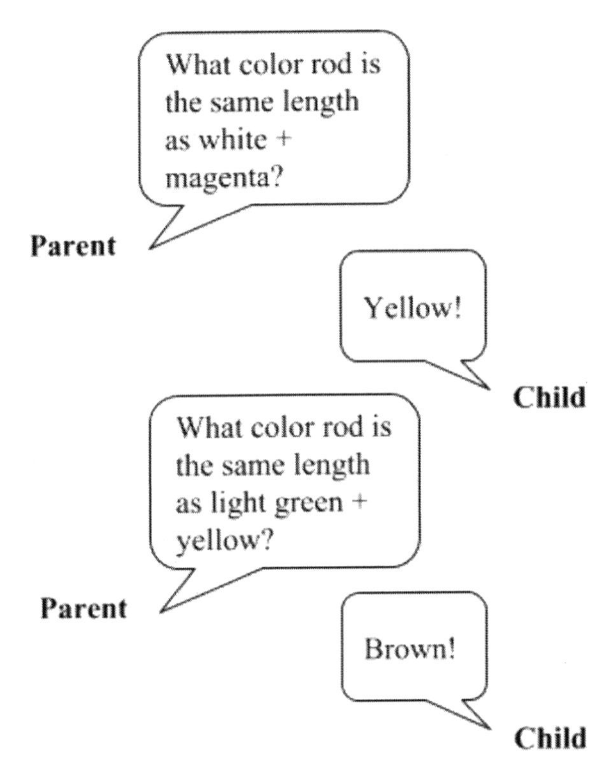

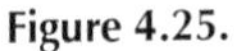
Figure 4.25.

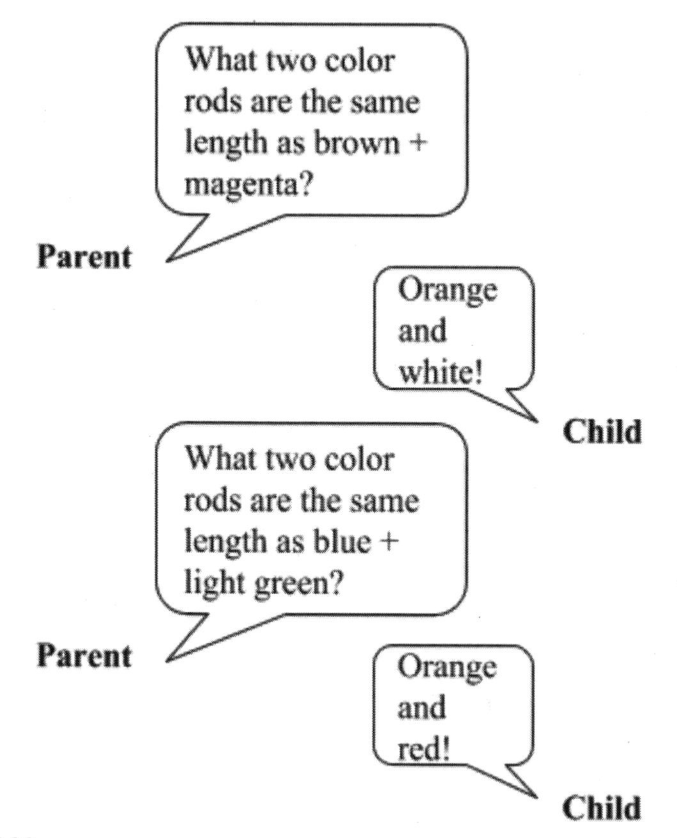

Figure 4.26.

This is a fun activity for kids and also a great activity to help kids see that, for example, 9 + 3 = 10 + 2.

Grades 3–5

Cuisenaire Rods are by far one of the best manipulatives available for developing number sense with fractions. Here, we will share one of our favorite activities using the rods, but remember, there are tons more! What we really like about Cuisenaire Rods is that the rods are not assigned a value, unlike base-ten blocks. While the rod is 10 centimeters long, it is not always 10 units long. In fact, when it comes to fractions, you can make any of the rods to represent a whole unit! Look at the example in figure 4.27 to see what we mean.

When you are ready to increase the rigor, change the whole from the orange rod to a different rod! (See figure 4.28.)

Another way to increase the difficulty is to change the structure of your question (see figure 4.29).

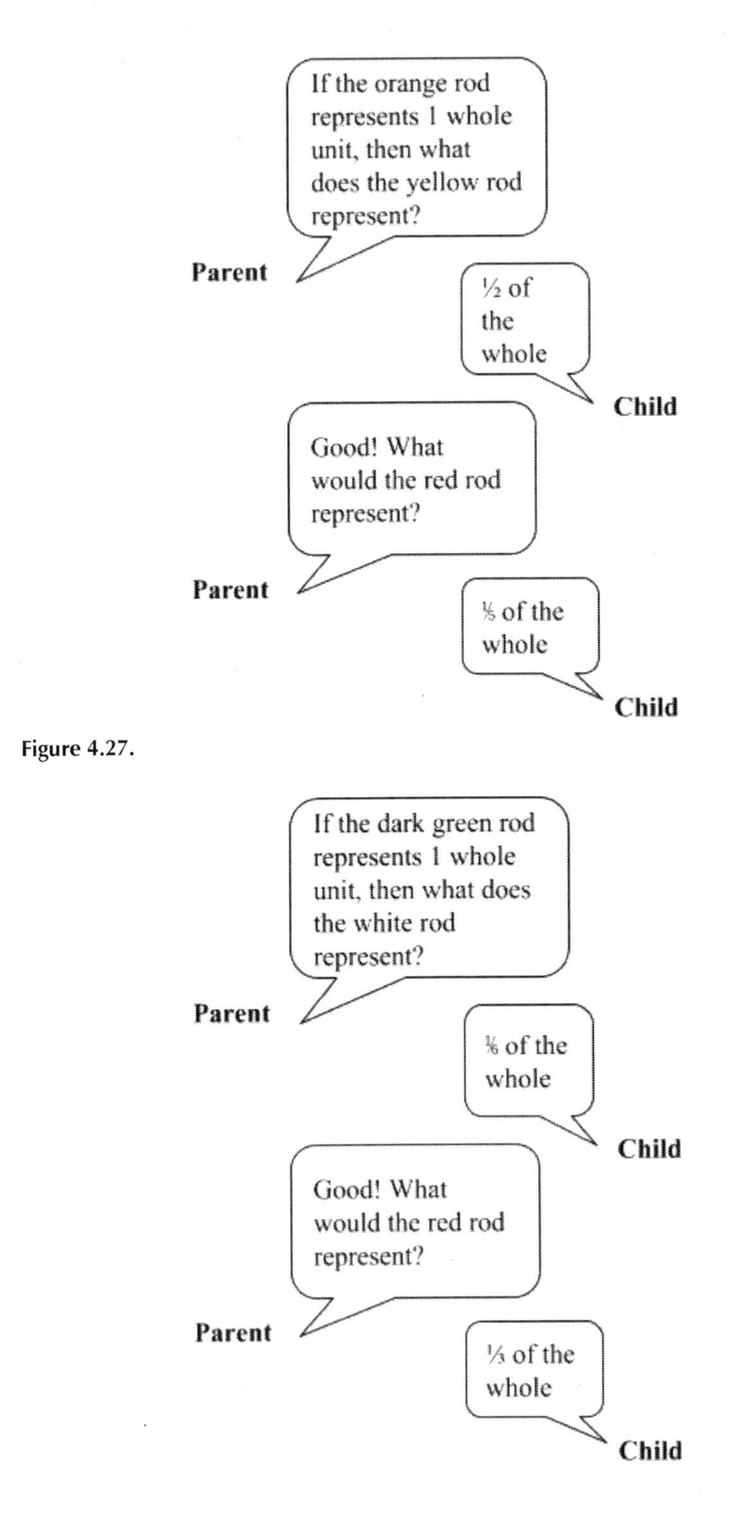

Figure 4.27.

Figure 4.28.

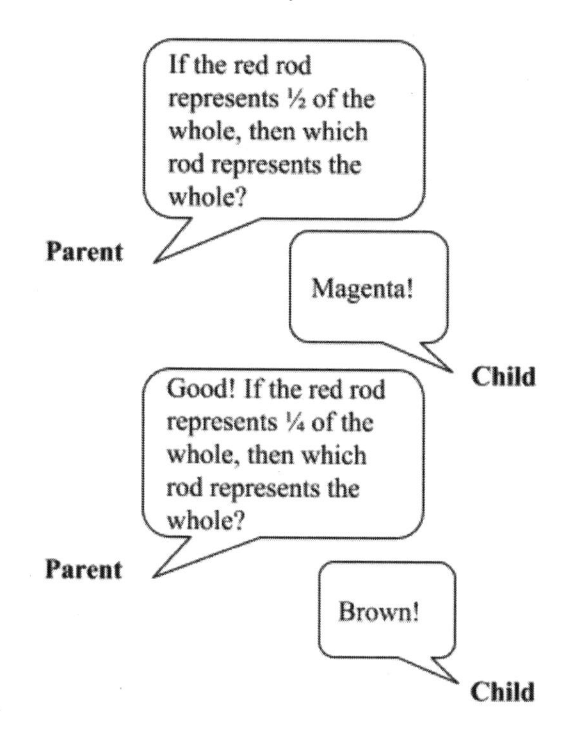

Figure 4.29.

Beyond Elementary School

Once your child has become fluent with Cuisenaire Rods and is able to identify fractional relationships, just as they did with fractions, continue with percents (e.g., instead of saying one-quarter of a rod you'd say 25 percent). Another way to use Cuisenaire Rods beyond elementary school is for developing understanding with ratios, proportional reasoning, and modular arithmetic.

REPRESENTATIONAL MODELS

As you saw earlier, physical tools are classified as manipulatives, or concrete, hands-on models. Because kids can manipulate them with their hands, they act as a great starting tool for many mathematical concepts. Referring back to chapter 1, many children learn mathematics best through a CRA approach. Tools are a great way to help support student learning through the concrete stage of learning new math concepts (at *any* age!). As children become more comfortable with certain math concepts through the use of manipulatives, they then should start to represent those hands-on tools by drawing pictures of them.

Representational models are tools, too, but rather than being tangible like manipulatives, they are *representational*, or drawn. It's important to note that representations are *not* exact copies. This is important because young children can sometimes get wrapped up in drawing perfect pictures that they lose track of the actual math concept. For example, if given a word problem such as, "Jimmy had 10 cookies. Joanna gave him 7 more. How many does he have now?" in our experiences, students not taught the difference between drawing a picture and drawing a representation will spend significant time drawing 10 perfect cookies (usually chocolate chip—we can tell by the precise dots inside the ovals) rather than just *representing* cookies with a circle or other symbol. This is a great place for parents to jump in and support their child's learning without worrying about teaching the math content incorrectly.

Below we've listed questions you can ask to better support your child through drawing in math.

- *Can you draw a picture to represent the situation or story?*
- *Where in your picture can I see the quantities from the story?*
- *How does your picture represent the story?*
- *How can you use your picture to help you make sense of the situation?*

Both manipulatives and representations will help your child build a deeper understanding of numbers, and ultimately, they will use their number sense to reason and think mathematically, especially when confronted with a problem they have never seen before. They will also be able to extend their thinking to the basic operations of addition, subtraction, multiplication, and division, which we will be exploring in the next two chapters. This foundation is critical for mathematical fluency and development.

Chapter 5

Understanding Whole Number Addition and Subtraction

> "Memorized knowledge is knowledge that can be forgotten. Internalized knowledge can't be forgotten because it is a part of the way we see the world. Children who memorize addition and subtraction facts often forget what they have learned. On the other hand, children who have internalized a concept or relationship can't forget it; they know it has to be that way because of a whole network of relationships and interrelationships that they have discovered and constructed in their minds."
>
> —Kathy Richardson

In our constant interactions with parents, we find they are very concerned about their children not learning the math they once learned and, specifically, using procedures such as "carrying" and "borrowing" in addition and subtraction. As our good friend and mathematician at large, James Tanton, frequently says, "adults equate familiarity with understanding." In other words, "I know it; therefore, I understand it." Just because one knows something, doesn't mean they understand it.

This chapter is meant to illustrate the various ways one could add or subtract and still arrive at the same answer as when done traditionally, using the U.S. Standard Algorithms. When we say standard algorithms in regard to addition or subtraction, we are referring to the way most American parents of today learned how to add and subtract when they were in school (see figure 5.1).

Before panic mode ensues, we assure you that your child *will* still learn how to add and subtract using the standard algorithms you once learned, or variations of them. However, they will learn them later in their academic career than when you did. Most students will master the standard addition and subtraction algorithms in fourth grade.

Figure 5.1.

The idea behind this is that many adults today cannot actually *explain* what they are doing when they operate with numbers. Instead, they follow procedures, or steps, and focus on getting an accurate answer. Children today will grow up in a world where machines and calculators can operate with numbers more accurately than humans and so they need to be able to *reason* through tough problems and *analyze* others' work rather than focus on the answers.

To get kids to *think* about the math, we have to give them opportunities to make sense of it. Once someone learns a fast and efficient method for solving a problem, it's hard to work backward and explain *why* it works. So, instead, we teach *why* first, allowing kids to explore the math and determine rules and procedures for themselves. Gone are the days when teachers hold all the knowledge and give it to their students—today we value the knowledge students already bring to school and use that to stretch their thinking.

Many parents have told us that they feel their children are speaking another language when it comes to the math they are learning today. Throughout these chapters, we hope to clarify some of that language for you. In terms of addition and subtraction, a major shift parents might notice is their child saying "regrouping" and "ungrouping" or "bundling" and "unbundling" instead of "carrying" and "borrowing." This is not done to annoy parents and adults, nor to make things more complicated. The reasoning is actually fairly simple: it's more mathematically accurate!

The words *carry* and *borrow* have alternative meanings in the English language, and using words with multiple meanings can impact a child's ability to understand the mathematics. In our daily use, the word *borrow* means you are taking something from someone with an intention of returning it. In subtraction, we don't return the "borrowed" unit, so this makes the process confusing for children.

Additionally, *carry* and *borrow* are not related terms. Because addition and subtraction are so related, it's important that we use terms that show the relationship. When we regroup, we are rearranging the current structure without changing the value (just grouping it a different way!). When we ungroup, we undo that rearrangement. As you progress through this chapter, take note of the various strategies for adding and subtracting and think about how the terms *regroup* and *ungroup*, or *bundle* and *unbundle*, are more applicable than *carry* and *borrow*.

As we saw in chapter 1, the skills needed to be successful in the future require much more than being a quick calculator. In this chapter, we will break down some common addition and subtraction techniques kids are learning today. *Note: The goal is not quantity of strategies but, rather, quality of understanding.*

It's not expected that your child will be able to use each strategy, or model, but rather that they should be exposed to many different strategies and models so they can choose the techniques that make the most sense to them. Realistically, even back in our day, we didn't ever expect cashiers to pull out pencil and paper to do long subtraction to make change for us, which for most of us was the technique we were taught in school, but, rather, to do the natural thing with numbers in that context: make sense of numbers. Today, the plethora of different contexts in the world means we need to teach students to be flexible with numbers and choose strategies that work for them.

EQUALITY IN MATHEMATICS

Before learning addition and subtraction, students should have a strong understanding of what the math term *equals* (=) means. Many students who lack this understanding begin learning how to operate and inevitably interpret = as "the answer" or as a directional symbol for where to write the answer. If your child does not exhibit the conceptual knowledge that the equals sign represents balance, or that the expressions on both sides of the sign hold the same value, then you should work with your child (and talk with their teacher) to build a more solid understanding before moving on.

One way to explain equality to a child is to actually use a pan balance. You can get these at any learning store or order one online. You may be thinking, "I can just draw a picture to show it." You are also right! But we caution you to always think back to the typical learning trajectory of mathematics (Concrete-Representational-Abstract, CRA). Start concretely, then represent it, then turn it into a symbol or abstraction.

We suggest taking an actual pan balance and filling one side with four cubes (be sure they are the same type of cube for weight equality). Ask your child to make it so that the balance is level. They will then need to fill the other side with the same number of cubes. Once they level the balance, have your child explain why the balance is level. If your child responds with, "both sides are the same" we encourage you to push back and ask them to be precise. Not all balanced situations have the same exact objects or numbers on both sides. For example, maybe your child put two red cubes and two blue cubes on one side and one yellow cube and three green cubes on the other side. The values are the same, but the expression is different!

Following the trajectory for learning math, have your child draw a picture to represent the situation. Finally, ask them to write a number sentence (equation) that mathematically tells what the picture shows. Once they've done this, point to the equal sign and ask, "What does this mean?" Continue to help your child build a deep understanding of equality before rushing them into learning how to operate.

THE OPERATIONS OF ADDITION AND SUBTRACTION

To start, we want you to take a moment to think deeply about the following questions. You may even want to write down your thoughts so you can refer back to your original thinking.

1. How would you define the operation of addition?
2. How would you define the operation of subtraction?
3. Does addition always result in more?
4. Does subtraction always result in less?

What Is Addition?

So, how did you answer the first question? How would you define the operation of addition in mathematics? If you said or thought something along the lines of "addition is the mathematical process of combining two or more quantities together," then you can pat yourself on the back. Addition is represented mathematically by the symbol "+" and can be interpreted as quantities joining or being put together. Kids learn from a very early age the mathematical language of addition. We recommend you using these terms as soon as your child starts preschool so you can model precise language (see figure 5.2).

The numbers or quantities we join are called the **addends**, and the total of the quantities is the **sum**. One of the expectations of math students today is to be able to communicate precisely. One way we encourage students to do this is by using the actual math language assigned. This helps students make more sense of the work they are doing and helps others interpret their thinking. If saying "sum" is too confusing for your child because of its oral similarity to "some," feel free to say "total."

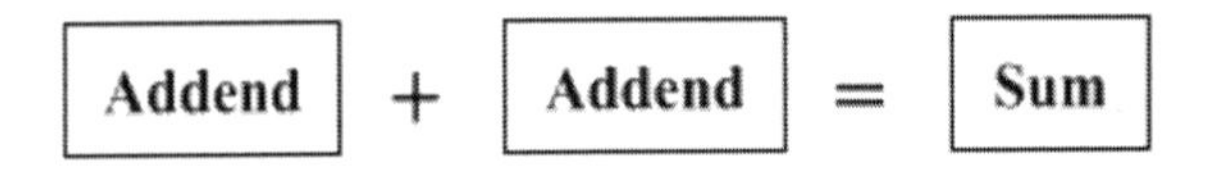

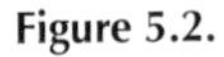
Figure 5.2.

TIP

It may assist you to refresh your memory of the properties of numbers that involve addition as you continue to read this chapter. Though there is no expectation for students below grade 3 to learn the names of the properties, as a parent, knowing the properties and how they work will better help you support your child. The properties are the commutative, associative, and additive identity properties.

The *commutative property of addition* tells us that the order of the addends does not matter in finding the sum. For instance, if you had three red blocks and two blue blocks, together you'd have five blocks. If you started the count with the two blue blocks first and then added the three red blocks, you'd still have five blocks. Alternatively written, $3 + 2 = 2 + 3$. Sometimes young children call this the "flip-flop" property. It's okay for kids to use this terminology, but remember, our goal is to communicate precisely in math, so if your child says that, respond by saying, "That's right! That's what we call the commutative property of addition because the order doesn't matter!"

The *associative property of addition* tells us that the sum is the same regardless how the addends are grouped. For instance, $(2 + 3) + 1 = 2 + (3 + 1)$. Notice how we did *not* change the order of the numerals. All we did was relocate the parentheses to a different grouping. By grouping the parentheses differently, we are really saying $5 + 1 = 2 + 4$.

The *additive identity property* tells us that any time we add zero to a quantity, the sum remains unchanged. For example, $4 + 0 = 4$. When you hear the word *identity*, think of it defining who a person is. In this case, it's defining what the sum is. Adding zero to the original number did not change the sum (its identity remained the same).

Returning back to what addition really means, now we want you to think about if, when adding, the sum is always greater than the addends. Many children are taught from a very young age that when they add numbers together, they will get a bigger number. One of the areas of our work is in helping teachers and parents understand that these general statements may work when children are young, but they eventually expire. For example, in kindergarten through sixth grade, kids will only experience adding whole numbers. When two whole numbers greater than 0 are added together, they will always result in a greater sum.

But, what happens when kids add 0 to a number? The result is not greater! Or, better yet, what happens when kids move on to seventh grade and they start operating with **integers**, or positive and negative whole numbers? The result won't always be greater. For example, $7 + (-3) = 4$. This is why we want you to define addition as the joining of two or more quantities and have a strong foundational understanding before working with your youngster.

What Is Subtraction?

The inverse operation of addition is subtraction. This means that it is the opposite of addition—namely, subtraction is not the joining of quantities but, rather, finding the difference between two quantities. Subtraction is represented mathematically by the symbol "–" and can be interpreted in three manners: (1) take away; (2) the distance between two things; or (3) comparison. Just like with addition, we encourage teachers and parents to be precise with their mathematical language. Similar to the parts of an addition equation, in math, we have specific parts of a subtraction equation (see figure 5.3).

The **minuend** is the number that will be diminished, or reduced. The **subtrahend** is the number to be subtracted from the whole. The **difference** is the result of subtracting the subtrahend from the minuend, or the missing part.

Did You Know?

The word *subtraction* is an English word derived from the Latin word *subtrahere*, which is a combination of *sub* (from under) and *trahere* (to pull). Put it together and what have you got? Subtraction is to take from below and remove.

Unlike addition, subtraction is not commutative—this means that changing the order *does* impact the difference. For example, $6-3 \neq 3-6$. Also, subtraction is not associative—meaning that one cannot regroup the quantities and get the same result. For example, $(6-3)-2 \neq 6-(3-2)$ because the left side is equal to 1, while the right side equals 5. Children begin to realize these properties through repeated exposure to story contexts in which they notice the math in a real-life situation.

Similar to addition, many children are taught from a very young age that when they subtract numbers, the difference will be smaller. Using arguments

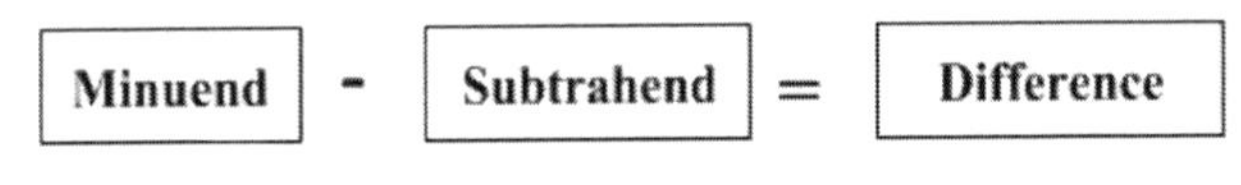

Figure 5.3.

similar to the ones we used for addition, it is critical that both teachers and parents explain concepts in ways that don't compromise future applications.

For example, in kindergarten through sixth grade, kids will only experience subtracting whole numbers. When two whole numbers greater than 0 are subtracted, they will always result in a smaller difference. But, what happens when kids subtract 0 from a number? The result is not less! Furthermore, what happens when kids move on to seventh grade and they start operating with integers, or positive and negative whole numbers? The result won't always be less. For example, $7 - (-4) = 11$.

Many adults learned about addition separately from subtraction when they were in school. Because addition and subtraction are so related, it is more advantageous to teach them *simultaneously*, rather than in isolation. In fact, once students begin to see the relationship between these operations, they realize they have fewer basic facts to learn because they can use the related operation to derive an answer. As we progress throughout this chapter, you will see us use both addition and subtraction to solve problems. We encourage you to let your youngster, when ready, explore the two operations in conjunction with each other.

Before we introduce various ways students are currently learning how to add and subtract, we want to highlight again the importance behind *conceptual understanding* versus *procedural understanding* to help you see why we are stressing the importance of thinking mathematically.

Imagine the following situation: A teacher writes the subtraction problem $13 - 7 = ?$ on the board, horizontally. Students are asked to find the difference. Conceptual thinkers may use addition to help them solve it $(7 + \square = 13)$ or they may use other facts they know, such as $10 - 7 = 3$ and then add back the 3 from the 13 to solve, or they might use a **doubles fact**, such as $7 + 7 = 14$, and then take away 1 to get 13. Other students may just know their addition and subtraction facts to 20 and can instantly spew out "6."

Figure 5.4 shows how a procedural thinker would solve the problem (and we have really seen this countless times!).

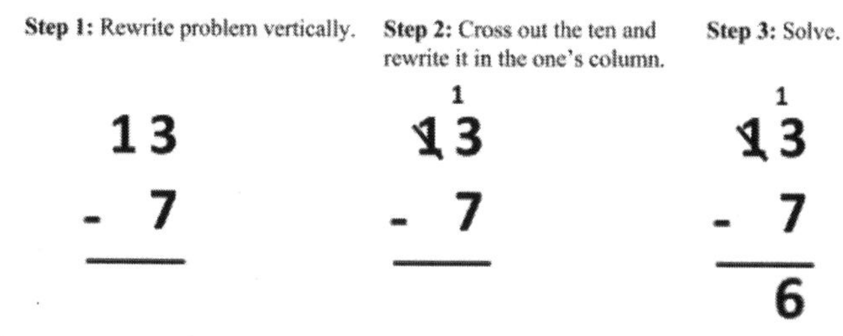

Figure 5.4.

$$1,002 - 998 = \square$$

Figure 5.5.

How did the student's procedural understanding help them solve the problem? In essence, all they did was cross out a 1 and rewrite it. Nothing actually changed! They still have to solve $13 - 7$. When students solve problems such as this trying to use procedures they don't understand, they waste time and struggle to apply what they do know to a problem.

Another common issue we see with procedural thinkers is that they don't know when and when not to use the procedural strategy. The goal of using the standard algorithm is to make the math calculation quick and efficient. But, does it always do that?

Try to solve the problem in figure 5.5 as quickly as you can.

We would stand to bet that you were able to solve this quicker with a mental math strategy than with pen and paper using the standard algorithm. Do you see that procedures aren't always more effective or efficient? If not, then pay close attention to the various strategies we offer throughout this chapter and come back to this section later.

We realize that, at first, some of the strategies we show you may look complicated. It's easy for adults to quickly dismiss the strategies because it seems like more work with fewer benefits than using the standard algorithm. However, we encourage you to try your best to notice within the strategies what conceptual understanding might be helpful and applicable later on when a student does learn the standard algorithm. Think about when each of the strategies presented might be most efficient or effective for particular problems. We will warn you, though . . . you may find yourself in the grocery store using your new strategies, rather than drawing the standard algorithm in the air!

Preschool–Grade 1

At the youngest level, children are often exhibiting early numeracy skills and are still working on solidifying counting with cardinality, recognition that the last number stated when counting objects represents the number in that set. Depending upon your child's counting abilities, we present various activities you could do at home to support their growth.

In figure 5.6, look at the hands holding up eight fingers. In this particular image, the child has three fingers visible on the hand on the left and five fingers visible on the hand on the right. Using this image, we are going to outline

Figure 5.6.

the three stages of a counting child. Try to identify what type of counting your child would use to find the total number of fingers visible on the hands.

Count all: The earliest stage of counting is the student who counts all. We call it that because this student needs to literally count each finger (or object). A child who is counting all would count all of the fingers above and successfully identify how many fingers there are in total. This means a child at this stage has cardinality. The count would sound like this: "1, 2, 3, 4, 5, 6, 7, 8."

Count on from the first number: Children who are at this stage will recognize that they don't need to count each and every finger because they can "hold" the first number in their head. This child is able to subitize and see that the left hand has three fingers without counting each of them individually. *Reminder: children don't need to know the word* subitize*—but we want to convey the concepts precisely to you as adults.* A child who can count on from the beginning quantity will sound like this: "3 . . . [pause], 4, 5, 6, 7, 8." The child will know to stop because they have *counted on* five more numbers (to represent the five fingers on the second hand in figure 5.6).

Count on from the greater number: A more efficient method is to count on from the greater number. A child at this stage will recognize that counting by ones, or counting all, takes too long and that counting on from the first number (in this case, the left hand—3), is not as fast as starting with the greater number (in this case, the right hand—5) and counting on from 5. A child using this strategy will sound like, "5 . . . [pause], 6, 7, 8." They will know when to stop because they have *counted on* three more numbers (to represent the three fingers on the first hand). Children at this stage are making use of the commutative property of addition because they recognize they can add the addends in any order.

When thinking about your child, what type of additive counter are they? If your child is not counting on from the greater number, do not panic! We advise you *not* just to teach your child to do that, as then you will have taught a procedure without the conceptual understanding component necessary for them to use and apply that strategy in other situations. Instead, give your child multiple opportunities to repeat their strategy until they realize there has to be a quicker way to solve the task.

Once your child has repeatedly counted on from the first addend, they will begin to notice that there are easier ways on their own. Letting your child come to the realization on their own that there are more efficient strategies is exactly how you want to facilitate their learning. Being patient is the greatest advice we can offer you.

How to Help at Home

For the "non-counting-on" child: Practice counting forward and backward from numbers other than 1 and 10. This will help your child begin to make sense of the order of our number system and recognize that they do not always need to start at 1 to count forward. Your child may need a tool to help them see the order of numbers. You may want to make a number path for them (you can do this by making your own version of the image in figure 5.7 for the numbers 1–100 and taping them together). If you make your own version, we suggest coloring each multiple of 10 (e.g., 10, 20, 30, etc.) to help your child learn to count by tens and to identify the beginning of each new decade.

Ask things like, "Can you start at 14 and count by ones for me?" If your child doesn't understand the task, rephrase your question: "What number comes after 14? 15? 16? Good. Keep going on your own until I say stop." Have them count until 32 so they can practice crossing over a decade (a multiple of 10).

Remember, anything you do forward, you should do backward. So, you can then ask, "Can you start at 32 and count backward for me until I say stop?" Interestingly, you will find your child is more than likely slower at counting backward than forward. This is a very hard task for children learning how to count, so be sure to be patient and keep providing the opportunities for them to practice the skill!

1	2	3	4	5	6	7	8	9	10	11	12

Figure 5.7.

To help get your child ready for counting on, you also want to practice skip counting, or counting by more than ones. Refer back to chapter 4, where we described how to teach your child to skip count by two. You also want to practice skip counting forward and backward by 3, 4, and 5, and you can do this using the same technique as we did with counting by twos. Don't forget to practice skip counting forward and backward by 10, both on and off the decade. Look at figure 5.8 to see the difference.

For the "counting on" child: If your child is already counting on, then you will want to help your child begin to identify more efficient strategies and use the structure of our number system to help them make sense of problems. Many children who are using counting on as a strategy rely on their fingers to count. This is a great strategy, and we encourage you to allow your youngster to use their fingers in math (it's even okay for you, as an adult, to use your fingers . . . really!).

Once students are comfortable with their fingers and counting on, they often can be very successful with problems such as 8 + 13 or 12 + 9, so long as they keep track of their count, because they realize that addition is commutative and that they can start with the greater number (to count fewer times) and count on. However, when given a problem like 12 + 13, many children will attempt the problem using their fingers and then say, "I don't have enough fingers" and give up.

If your child is at this point, ask guiding questions that lead them toward identifying a solution. For example, "You are right; you don't have that many fingers. Could you keep track of your count with something other than fingers?" Guiding your child to identify that when a challenge is presented, finding other solution paths can be helpful is a skill that will be useful long after elementary school. If they struggle, help them look around the room for other things that they could use to keep track of their count, such as windows, doors, floor tiles. It doesn't matter if they count 10 fingers and 2 doors because their count of 12 is still being tracked.

Counting forward →

On the Decade	10	20	30	40	50	60	70
Off the Decade	13	23	33	43	53	63	73

← Counting backward

Figure 5.8.

This is where the structure of numbers becomes really helpful for a counting on child. Understanding *five-ness* and *ten-ness*—and yes, those are real words—is critical for understanding our number system. *Five-ness* and *ten-ness* are simply terms used to describe having a strong sense of those numbers. Students who have "five-ness" are able to use the number 5 to help them solve problems. Helping your child see the structure of 5, and later 10, will allow them to be more successful later on.

A tool commonly used in classrooms that helps students see five-ness and ten-ness is a **ten-frame**. A ten-frame is a two-by-five arrangement of objects (remember, on this page it's a picture, but you should first have kids actually play with one!) purposefully structured to help kids subitize, or see more than one object and recognize it as a set. Figure 5.9 shows an example of a ten-frame that has 10 objects.

You can buy premade ten-frames online, or you can make your own by using tape, markers, or any craft. We advise you at first to work with ten-frames that have the grid and then as you notice your child showing a stronger sense of 10, remove the grid and arrange the objects in a ten-frame way (see figure 5.10). This is what educators call *gradual release*. It means we provide all the help they need up front and slowly take away the aids until the child can do it successfully on their own. Think about gradual release like teaching a child to ride a bike and slowly removing training wheels.

Do you see the difference? By not having that grid in place, students are forced to *imagine* it, which is much harder than just putting a dot in a box! Let's look at a few examples so you can see how five-ness (and later on,

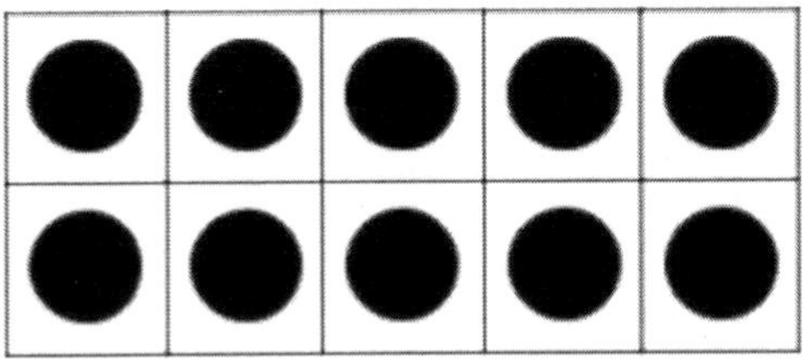

Figure 5.9.

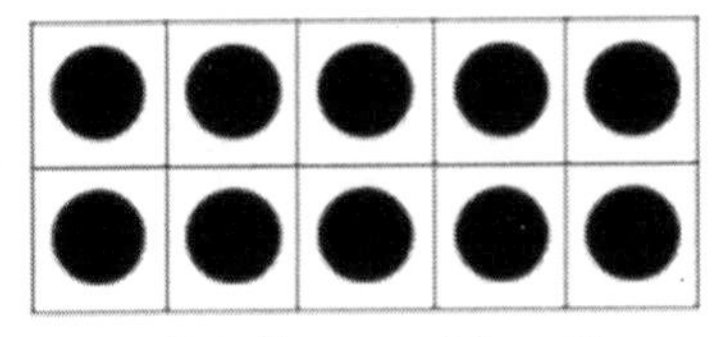

Ten frame with grid

Ten frame way

Figure 5.10.

ten-ness) is helpful. Look at the ten-frame in figure 5.11. What do you notice? How many dots do you see? How did you see them?

Students who are not using the structure of the ten-frame will often count by ones or sometimes by twos to find the total number of dots. But, what do you think children who have a strong sense of five see? They see five dots on the top row and three dots on the bottom row, and because they have *five-ness* they know there are eight dots altogether. Students could also say there are eight dots because they know a ten-frame has 10 boxes and two are empty, so $10 - 2 = 8$.

Children might also use their **doubles facts** to help them identify how many dots they see. For example, figure 5.12 has the same ten-frame shaded differently to highlight the doubles fact of $3 + 3$. A student may use the structure of the doubles fact to identify the base of six dots and then recognize there are two more, totaling eight.

By having these conversations with your child, you are helping them talk mathematically and see numbers as flexible things. In this case, an 8 is a $6 + 2$, or a $5 + 3$, or a $10 - 2$. Now, think about what this looks like once your child shows *five-ness*. How does *ten-ness* look? Figure 5.13 shows how a student might use ten-frames to see 10 as a unit.

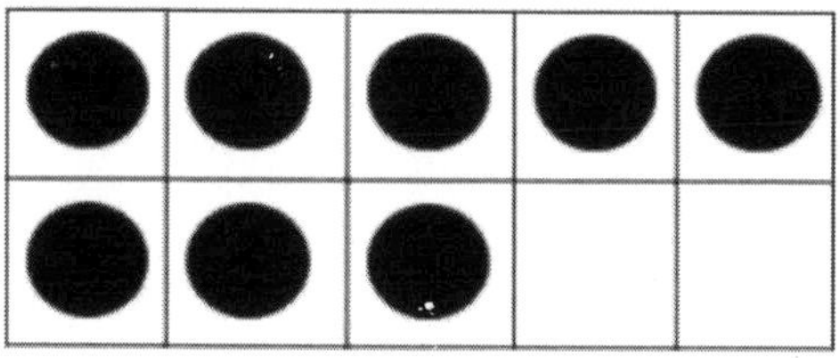

Figure 5.11.

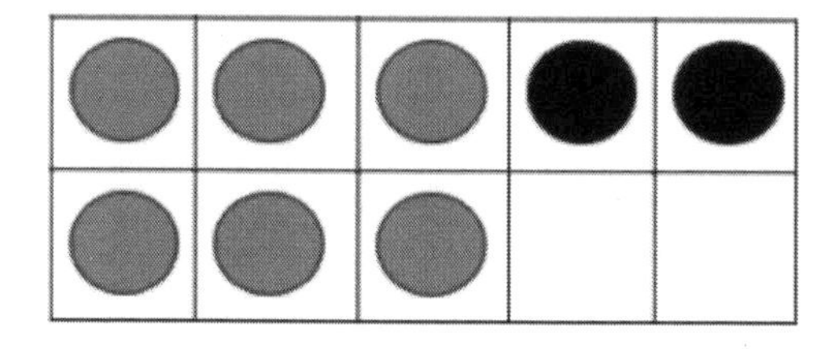

Figure 5.12.

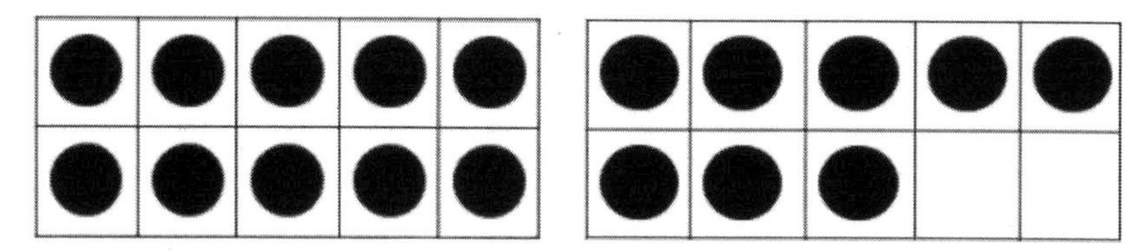

Figure 5.13.

Notice how the student can identify 18 because they know a filled ten-frame represents 10 and they knew 8 because of their strong sense of 5. Once students are familiar with ten-frames, you can help your child to use the structure of the ten-frame to add. For example, let's say your child is asked to add 8 + 5. As symbols, 8 + 5 is hard to visualize. So, you may ask your child, "Can you represent it on a ten-frame?" In this case, let's imagine figure 5.14 shows your child's representation.

What are some ways your child could use this ten-frame structure to help them solve the task? Are you picturing, perhaps, the first row of the left ten-frame and the first row of the right ten-frame combined, to make a 10? Then, there are just three left over! That is one way to think structurally about this. Figure 5.15 shows another visualization.

A student may notice that in the expression 8 + 5, the 5 could be **decomposed**, or broken apart, into 3 + 2. This is helpful because they can take two of the dots from the five dots and move them to the ten-frame with eight dots to make ten dots. The number 10 is a "friendly" number, or a landmark number, because it is easy to operate within our brains. Notice the complicated math that just took place: 8 + 5 was reorganized to show 8 + (2 + 3), which then (using the associative property of addition) looked like (8 + 2) + 3, or 10 + 3.

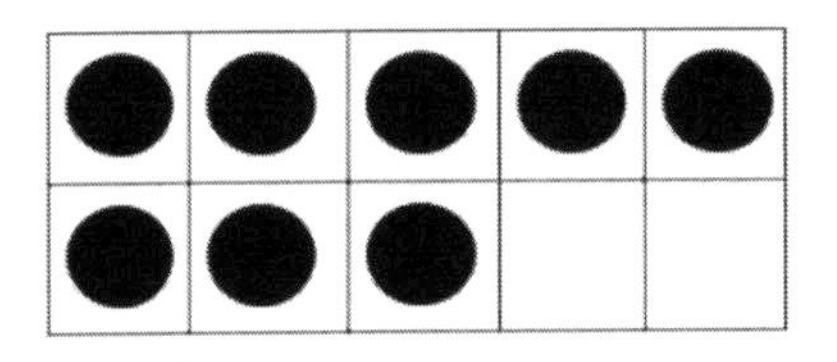

8 + 5

Figure 5.14.

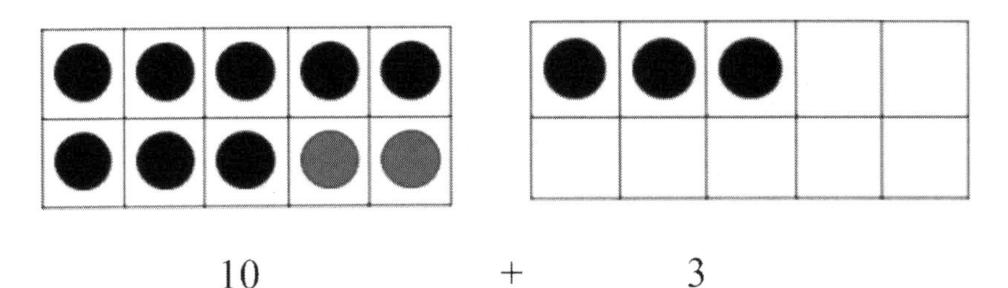

10 + 3

Figure 5.15.

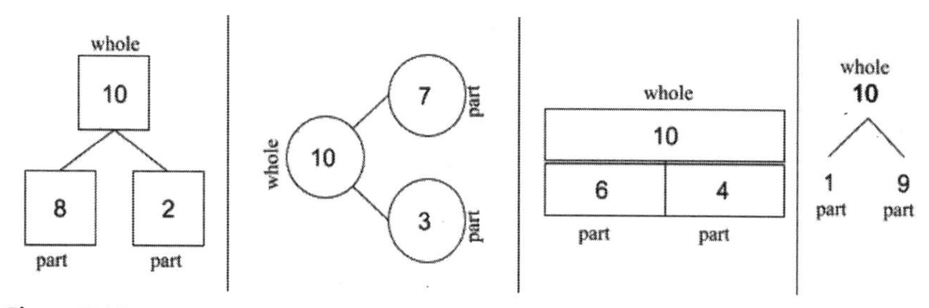

Figure 5.16.

Now, of course, a young child won't make the mathematical connection to the associative property yet, but *this* is why we teach math using manipulatives. It helps children see that numbers are flexible and later on helps them draw better connections to the properties of math that once were just a memorization drill for us adults.

Think back to chapter 3 (early numeracy) where we described the three main components of numbers: quantities, words, symbols. It is important that kids are exposed to numbers in many different ways, and ten-frames help them see the quantity component. Another tool that helps students see numbers in various ways is a **number bond**.

A number bond is a model that shows a quantity or numeral broken down to show a part-part-whole relationship. This model is helpful because it shows the relationship between addition and subtraction. Figure 5.16 shows examples of the way number bonds can be represented. You'll see that there are many shapes, structures, and orientations for a number bond. The essential part students need to know is that a number bond shows a whole and its parts—it doesn't matter if you use boxes, circles, or nothing to frame the numbers. What is evident from a number bond, though, is that the parts, when put together, compose the whole.

We particularly like the third representation of the number bond, as it leads nicely into **tape diagrams**, or **bar models**, which students at older grade levels use to problem solve.

STRATEGIES TO EXPLORE (GRADE 1 +)

Now that you have had a chance to see how number and operations develop early on, we will now model for you ways you can add and subtract with double-digit and triple-digit numbers. We do not intend for our lists to be exhaustive but, rather, to show you a variety of methods your child might be using or will use one day.

TIP

The way we present these is ***not*** *how they should be introduced or instructed to children.* We are showing *you*, an adult, the various strategies all at once to help you identify the method in which your child might be solving their problem and to give you some familiarity with it so you can talk with your child about their strategy. But, remember, these strategies are to be taught first through concrete manipulatives, then through drawings and representations, and finally through abstract symbols. In some cases, you will see us use a visual representation to help *you* understand the strategies, but all the strategies when taught to children should be done through the CRA approach. This instruction will take place in the classroom.

Your help at home in talking about math is all we can ask! Our greatest piece of advice is to not worry about *teaching* your child but focus on *communicating* with them mathematically.

Last, but not least, it is important that you recognize that these strategies are meant to build *mental* reasoning skills. They may not be the most efficient strategies, but in the beginning, we just want kids to be thinking mathematically and having several ways to solve one problem—without always having to write down their work. This will become useful when they are in situations, such as the grocery store, and don't have a pen and paper to use the standard algorithm, or their phone battery has died and can't rely on their calculator.

Addition

Some strategies you may see your youngster use to solve:

$$\mathbf{37} + \mathbf{83} = ?$$

1. Base-Ten Block Representations

Base-ten blocks, as you learned in chapter 4, are physical manipulatives children use to concretely understand mathematics. While we are drawing representations in this book, please keep in mind that, in the beginning, students will be using actual blocks before representing them on paper with drawings. Teachers tend to use place-value charts (T-charts that are labeled with the place values) as a "placemat" for the blocks.

In our representations, the ones units are represented with small squares, the tens with a rod that represents 10 units together, and the hundreds with a large square representing 10 rods together or 100 units together. Figures 5.17 and 5.18 examine how students would join the two quantities to find the sum. Think about how this process connects to the standard algorithm you learned growing up.

2. Chip Model

The chip model is very similar to the base-ten block strategy, but there is one main difference: size of the manipulative. Base-ten blocks are representational tools, meaning students can just look at the rod and know it represents 10 because of its shape and designation. Similarly, a student could look at the large square and just know it represents 100 because they've been taught that. The chip model is different because the tool is nonrepresentational, meaning that all chips are the same size, so students cannot know just by looking at a chip the value it represents (see figure 5.19). This is critical because it forces students to develop a deeper understanding of the place-value chart and how to use it to support their understanding.

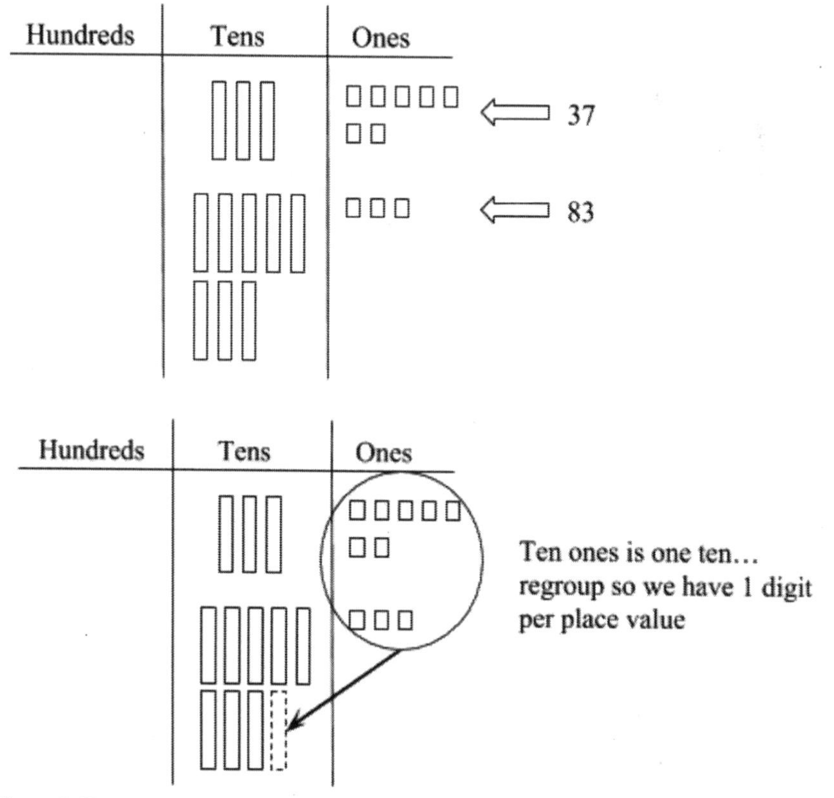

Figure 5.17.

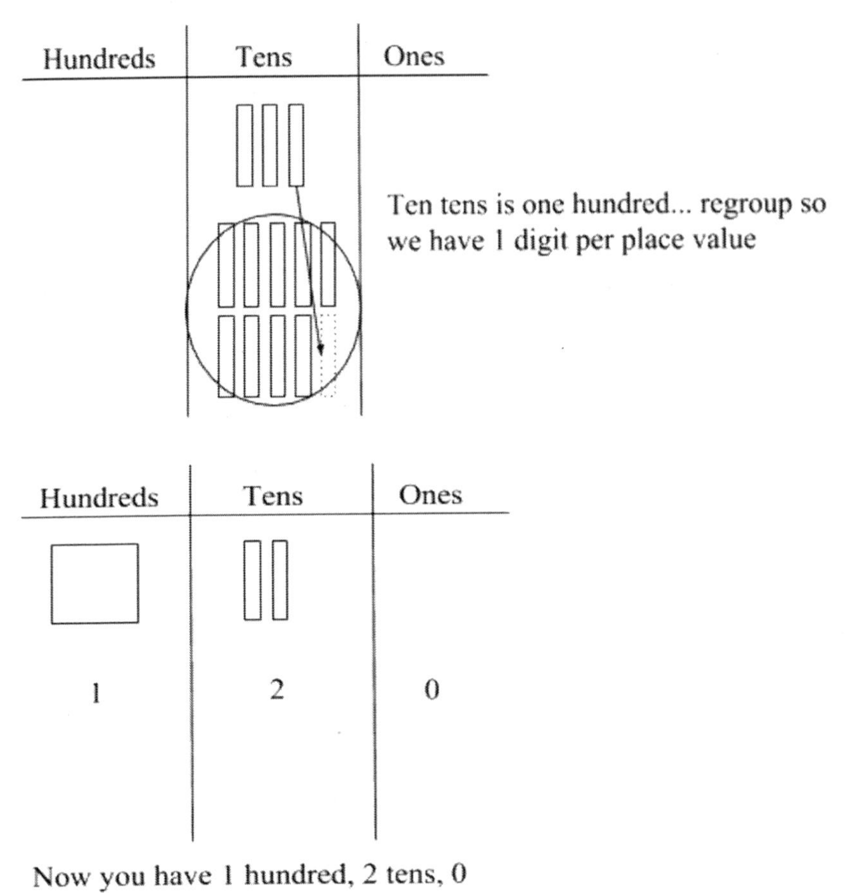

Figure 5.18.

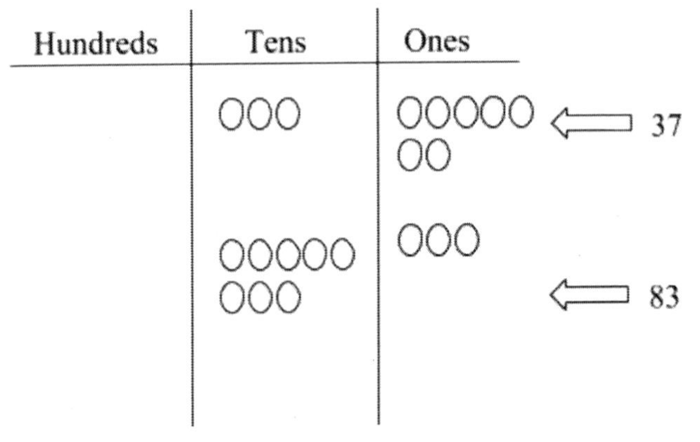

Figure 5.19.

Notice how we always organize our representations in the ten-frame way (five on the top, five on the bottom, in a two-by-five grid). This helps students recognize a quantity without having to count by ones. It also makes it easier for a student to see how to make a group of 10, which is important when we regroup so there is only one digit in each place value (see figure 5.20).

Hundreds	Tens	Ones
	OOO	OOOOO OO
	OOOOO OOO	OOO

Ten ones is one ten, so regroup so we only have one digit in each place value

Hundreds	Tens	Ones
	OOO	
	OOOOO OOOO	

Ten tens is one hundred, so regroup so we only have one digit in each place value

Hundreds	Tens	Ones
O	OO	
1	2	0

Figure 5.20.

3. Decomposition or Compensation

We can use a number line to help us break apart one of the addends to make the problem easier to solve. In this case, imagine a student started with 83, the greater addend, and then wanted to add 37. Using a number line, a child may "hop" to the nearest 10 and use that as their "friendly number." See figure 5.21 for a visual.

Your child might use decomposition without a number line. One way they could do that is by breaking apart the number 83 into 3 + 80. Figure 5.22 shows how they would write the decomposition to make it easy to see how to regroup the numbers so that they can make a "friendly" number, in this case, a multiple of 10. By breaking apart the 83 into 3 + 80, the child can then take the 3 from 83 and add it to the 37 to make 40. This helps a student solve the problem mentally.

Some might also call this compensation, since we added three to one addend to make up for the three we took away from the other.

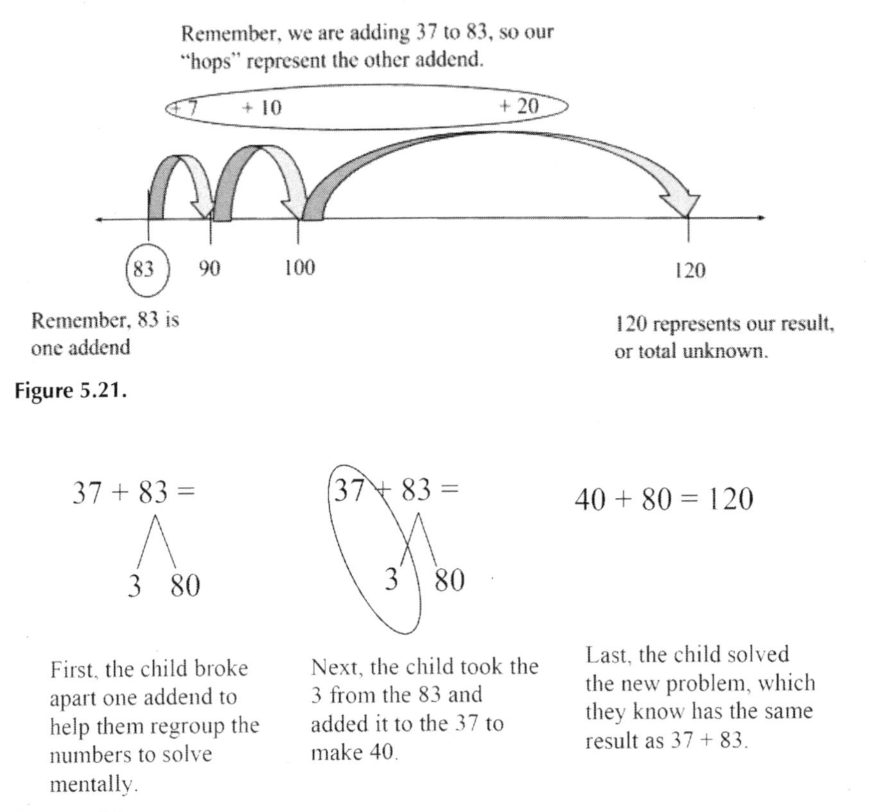

Figure 5.21.

Figure 5.22.

Another way your child could use decomposition to make the problem easier to solve mentally would be to break apart the other addend, 37, into 30 + 7. This would allow them once again to regroup the numbers to compose a "friendly number," in this case, 90, a multiple of 10. See figure 5.23 for a visual representation.

Some might also call this compensation, since we added seven to one addend to make up for the seven we took away from the other.

4. Partial Sums

In this strategy, students expand both addends and then add by place values. Mathematically, your child is just using the commutative property of addition by taking the **expanded forms**, or place values, of 37 and 83 (30 + 7 + 80 + 3) and reorganizing the expression so the place values align (30 + 80 + 7 + 3). Figure 5.24 gives a visual representation.

Children may write this strategy in a variety of ways, but we most often see it written in two ways: first, using the method shown in figure 5.24, and second, using the method shown in figure 5.25, which is closely aligned to the standard algorithm and helps children keep track of their work, while also developing a deeper understanding for the need to regroup.

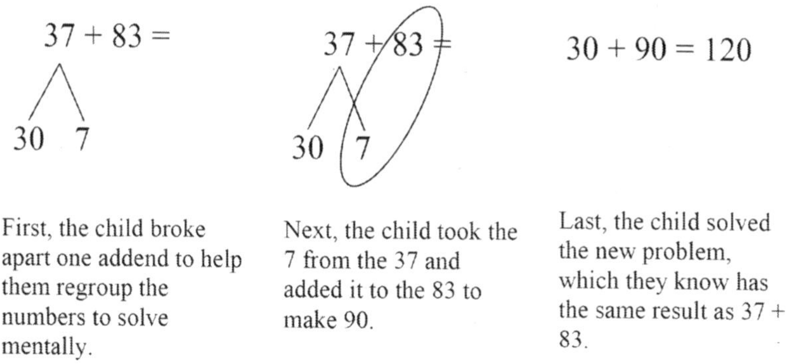

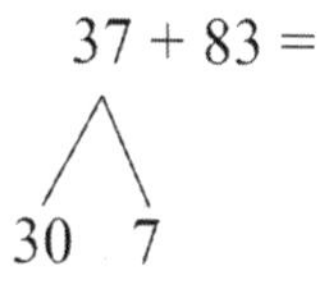

First, the child broke apart one addend to help them regroup the numbers to solve mentally.

37 + 83 =
30 7

Next, the child took the 7 from the 37 and added it to the 83 to make 90.

30 + 90 = 120

Last, the child solved the new problem, which they know has the same result as 37 + 83.

Figure 5.23.

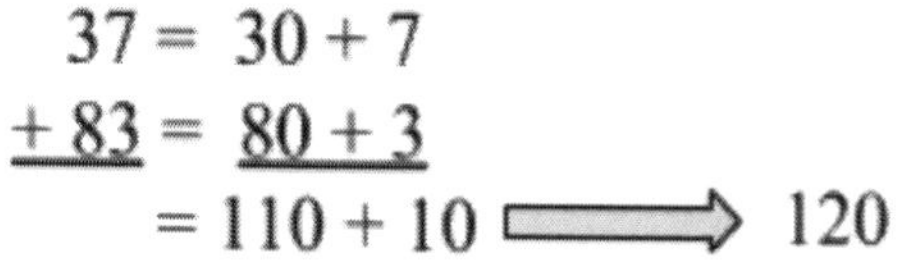

$$\begin{aligned} 37 &= 30 + 7 \\ \underline{+83} &= \underline{80 + 3} \\ &= 110 + 10 \Longrightarrow 120 \end{aligned}$$

Figure 5.24.

$$\begin{array}{rl} 37 & \\ +\underline{83} & \\ 10 & = 7 + 3 \quad \text{(add the ones)} \\ +\underline{110} & = 80 + 30 \quad \text{(add the tens)} \\ 120 & \end{array}$$

Figure 5.25.

So, how did you do? Do you think you could apply some of these strategies to another problem? Remember, children learn these strategies slowly and over years, building on foundational understandings. We just presented them to you in a matter of pages, so don't be ashamed if some of the strategies don't click right away. We suggest you try the next problem on your own, and then refer to your own child (if they are ready) to see how else they might solve it. Go ahead. *Get a piece of paper and try to solve 67 + 109 in as many different ways as you can.* We have provided suggested solutions on p. 98.

Subtraction

Some strategies you may see your youngster use to solve:

$$\mathbf{137 - 90} = ?$$

1. Relate to Addition

One way students learn to solve subtraction problems is to relate it to what they already know: addition. To solve this problem using addition, think about it as finding the distance between 90 and 137 on a number line (remember, all real numbers have *magnitude*, or a distance from zero, and can be placed on a number line). Figure 5.26 gives a visual representation.

Rather than thinking 137 – 90 = ?, the student would think, 90 + ? = 137. Students are then taught to jump to a "friendly" number. This terminology just means find a nice number that we can operate with mentally. In this case, many students would add 10 to 90 to get to the nearest hundred and then, if they have the ability, add 37 to 10 to obtain 47. If they are not ready to make that connection, they may just add 10, four times, and then add 7.

2. Decompose

Decompose in mathematics means to break apart. Usually, students will break apart a number by place values, but sometimes there are

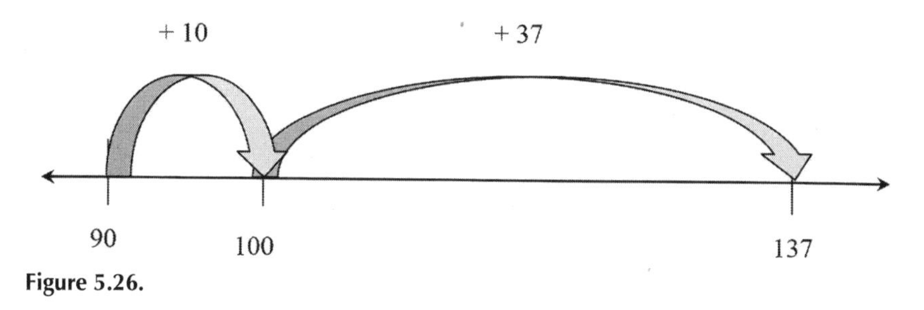

Figure 5.26.

137 - 90 =

130 7

First, the child ungrouped the minuend to help them subtract "friendly" numbers.

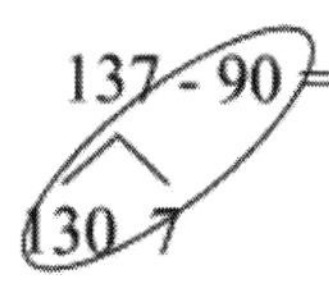

Next, the child subtracted 90 from 130, since they can do that mentally.

130 - 90 = 40
40 + 7 = 47

Last, the child added the remaining 7 to the difference between 130 and 90.

Figure 5.27.

circumstances in which that is not the most efficient or the simplest method to calculate mentally. Below, we show you four ways to use decomposing as a subtraction strategy for this one problem.

When choosing to decompose in this manner, you first might break apart the minuend into its place values to allow for easy mental computation (see figure 5.27). Notice, all we did was break apart 137 into 130 + 7. Ungrouping in this way helps you reorganize 130 + 7 − 90 to read as 130 − 90 + 7 because of the commutative property of addition. *Note: If a child is using this strategy but doesn't know 130 − 90, it would be wise to return back to reviewing basic facts, such as 13 − 9 = 4, in order to effectively apply prior knowledge.* By reordering the expression, one can solve 130 − 90 to get 40. Finally, add the leftover 7 to the 40 to obtain 47.

Unlike in figure 5.27, this time, in figure 5.28, we will decompose 137 into its **expanded form**, or place values (100 + 30 + 7). This allows us to visually see 100 + 30 + 7 − 90 and realize that we can reorganize the expression using the commutative property of addition (100 − 90 + 30 + 7). Then, 100 − 90 = 10, so 10 + 30 + 7 = 47.

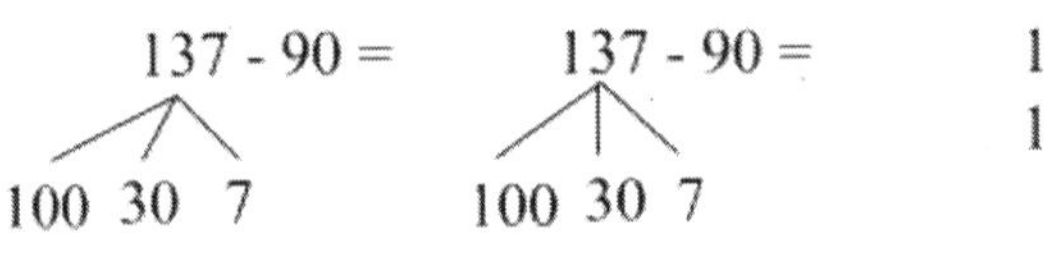

100 - 90 = 10
10 + 30 + 7 = 47

First, the child ungrouped the minuend to help them subtract "friendly" numbers.

Next, the child subtracted 90 from 100, since they can do that mentally.

Last, the child added the remaining 30 and 7 to the difference between 100 and 90.

Figure 5.28.

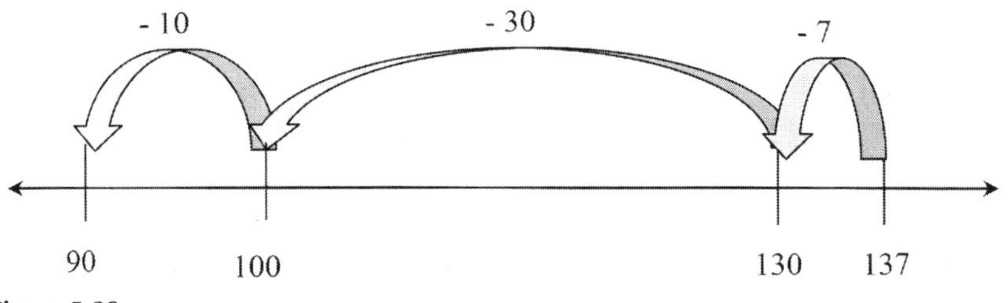

Figure 5.29.

Just as we jumped forward on the number line in relation to addition, we can also use the number line and decomposition to think about subtraction (see figure 5.29).

In this case, start on the number line at 137. For adults, you might be able to imagine subtracting 37, but for some students this may be tough, so we showed you how a student might decompose 37. Once you have subtracted 37 from 137, subtract another 10 more to get to 90. You've subtracted a total of 47.

As students become fluent with the properties of mathematics, they tend to use the structure of the math itself instead of a number line or other visual support. For example, we could rewrite the expression 137 – 90 as 137 – (7 + 30 + 50 + 3). Can you figure out why we would decompose 90 into 7 + 30 + 50 + 3? Well, we know 137 – 7 = 130. We also know 130 – 30 = 100. Further, 100 – 50 = 50. Finally, we can use that to solve 50 – 3 = 47. For this one, you could also use a number line to help you visualize.

3. Partial Differences

Partial differences is similar to partial sums. In this strategy, students expand the minuend and the subtrahend and then subtract by place values. See figure 5.30 for a visual.

In this particular problem, students who use this strategy would have to be comfortable with getting a negative number when they subtract 90 from 30. Most

$$\begin{array}{rcl} 137 & = & 100 + 30 + 7 \\ \underline{-\ 90} & = & \underline{\qquad\ -\ 90} \\ & & 100 - 60 + 7 \Longrightarrow 47 \end{array}$$

Figure 5.30.

students don't start operations with negative numbers until seventh grade, but kids who have a strong number sense in second grade can even use this strategy!

Subtraction Beyond Multiples of Ten

Some strategies you may see your youngster use to solve:

140 – 98 = ?

1. **Relate to addition**

   To solve this problem using addition, think about it as finding the distance between 98 and 140 on a number line. Instead of thinking about the problem as subtraction, think, "what can I add to 98 to get 140?" or, 98 + ? = 140. Figure 5.31 shows a visual representation.

2. **Decompose**

   Remember, there are many different ways to decompose. We show you a few here, but the idea again is that you are exposed to a variety of strategies so that you can communicate mathematically with your child.

   One way you might decompose this problem is by breaking apart the subtrahend (98) into its place values. See figure 5.32 for more details. Another way you might decompose this problem is by breaking apart the minuend (140) into its place values (see figure 5.33).

3. **Partial Differences**

   In this strategy, students expand the minuend and the subtrahend and then subtract by place values. Again, students need to realize that if they use this strategy, they will need to be able to recognize that they will be subtracting 90 from 40, resulting in −50. Kids who are not comfortable with subtracting a larger positive number from a smaller positive number should not choose this strategy. Figure 5.34 shows a visual representation.

4. **Overestimate**

   Sometimes, it is helpful to round a number up in order to get close to a "friendly" number; this aids in the mental math process. For a problem such as 140 − 98, rounding 98 up to 100 works nicely. Using the number

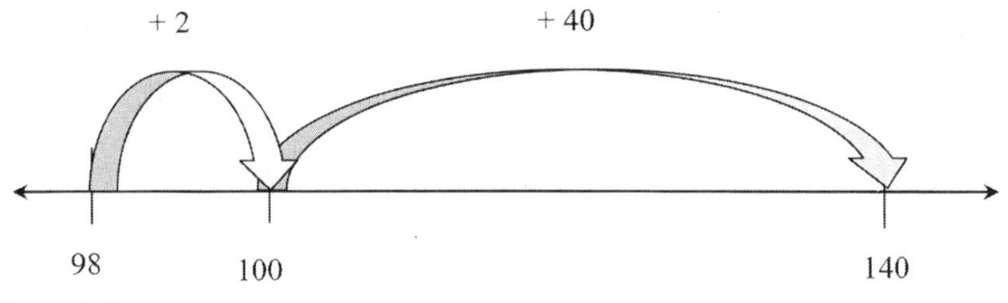

Figure 5.31.

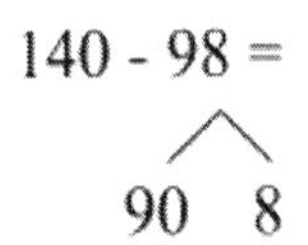

First, the child ungrouped the subtrahend to help them subtract 'friendly' numbers.

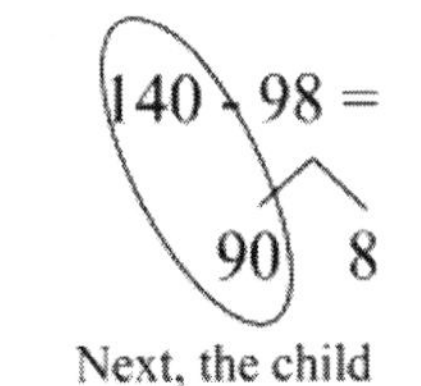

Next, the child subtracted 90 from 140, since they can do that mentally.

140 - 90 = 50
50 - 8 = 42

Last, the child subtracted the remaining 8 from the difference of 140 and 90.

Figure 5.32.

140 - 98 =
100 40

First, the child ungrouped the minuend to help them subtract 'friendly' numbers.

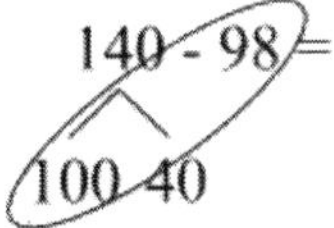

Next, the child subtracted 98 from 100, since they can do that mentally.

100 - 98 = 2
2 + 40 = 42

Last, the child added the remaining 40 to the difference of 100 and 98.

Figure 5.33.

1 4 0	=	100	+	40	+	0		
- 9 8	=			- 90	-	8		
		100	-	50	-	8	⟹	42

Figure 5.34.

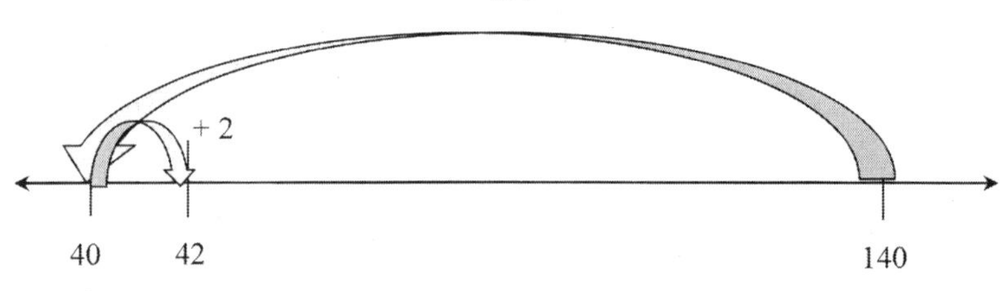

Figure 5.35.

line in figure 5.35, you can see that by subtracting 100, we are "over jumping" and, therefore, will need to add back the amount we over jumped. Can you see the subtraction of 98 within the number line?

5. Constant Difference

Constant difference is another method that can be helpful in certain situations. As it sounds, essentially you adjust the numbers of the problem so you will still have the same answer, but an easier problem to think about mentally. This strategy hones in on the understanding of equality. Constant difference works nicely for problems such as 140 − 98, where the minuend or subtrahend are close to a landmark number. Below, we show both the visual (see figure 5.36) and abstract representations of this method.

Visually

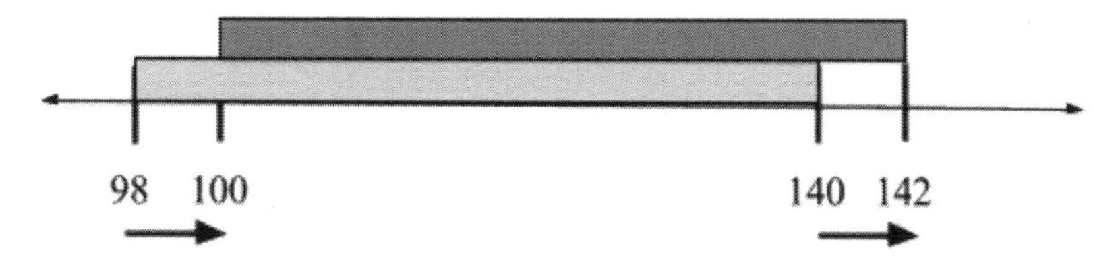

By adding 2 to each quantity in the expression, all we are doing is shifting the distance, not altering it.

Figure 5.36.

Abstractly

You can think of constant difference in this case as just adding two to both the minuend and subtrahend (140 + 2) – (98 + 2). By doing this, we've rewritten the problem to be 142 − 100, which is much easier to think about mentally than 140 – 98.

So, how did you do? Do you think you could apply some of these strategies to another problem? Remember, children learn some of these strategies slowly and over years, building on foundational understandings. Don't be ashamed if some of the strategies don't click right away—it is quite a challenge as an adult to "forget" all the procedures we already know and try to imagine learning from a child's perspective. We suggest you try the next problem on your own, and then refer to your own child (if they are ready) to see how else they might solve it.

Get a piece of paper and try to solve 181 − 49 in as many different ways as you can. We have provided suggested strategies beginning on p. 100.

ANSWER KEY FOR CHAPTER 5 ADDITION AND SUBTRACTION

Note: Answers will vary and can be written and represented in many different ways. These answers also do not represent an exhaustive list.

Addition: 67 + 109

1) **Base Ten** (see figure 5.37)
2) **Chip Model** (see figure 5.38)

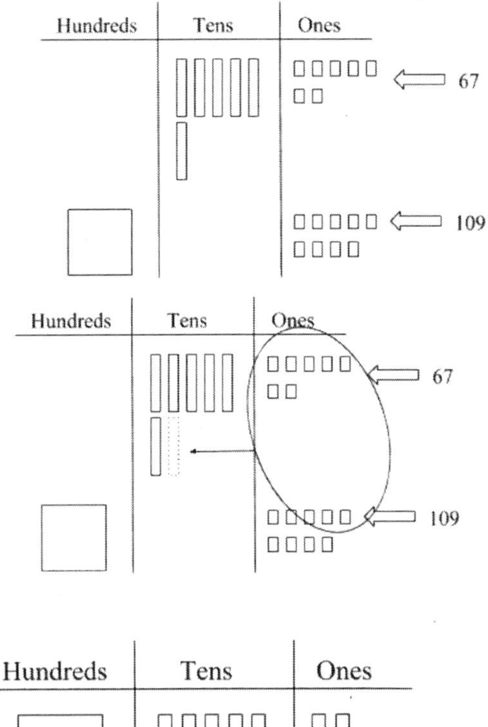

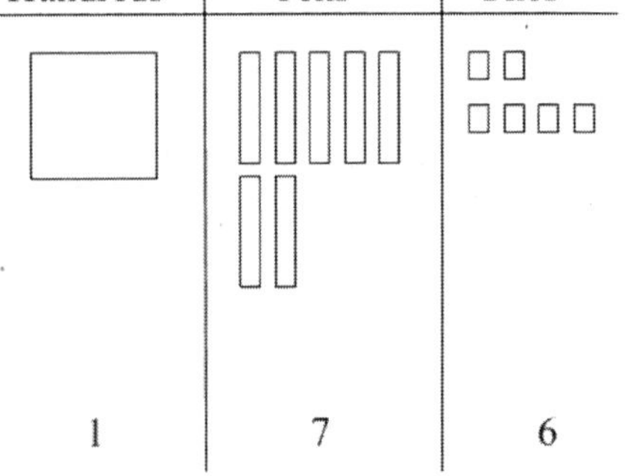

Figure 5.37.

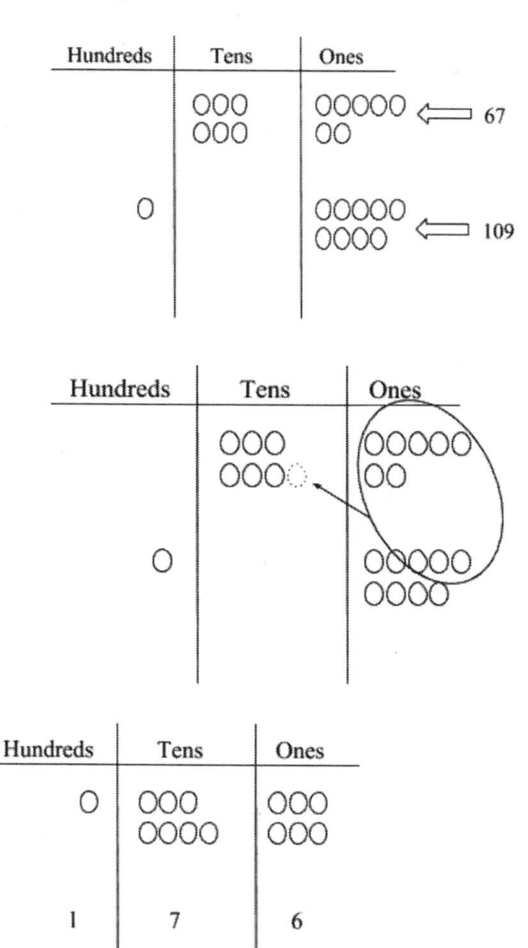

Figure 5.38.

3) Decomposition

One example of decomposition is shown in figure 5.39.

4) Compensation

$67 + 109 = (67 - 1) + (109 + 1) \rightarrow$ We can subtract one from one addend and add one to another to make friendlier numbers.

$66 + 110 = 176$

67 + 109 =
3 106
First, the child broke apart one addend to help them regroup the numbers to solve mentally.
67 + 109 =
3 106
Next, the child took the 3 from the 109 and added it to the 67 to make 70.
70 + 106 = 176
Last, the child solved the new problem, which they know has the same result as 67 + 109.

Figure 5.39.

$$\begin{array}{rcl} 67 & & \\ \underline{+109} & & \\ 100 & = & 0 + 100 \text{ (add the hundreds)} \\ 60 & = & 60 + 0 \text{ (add the tens)} \\ \underline{+\quad 16} & = & 7 + 9 \text{ (add the ones)} \\ 176 & & \end{array}$$

Figure 5.40.

132
+ 1
+ 50
+ 50
+ 31
49
50
100
150
181

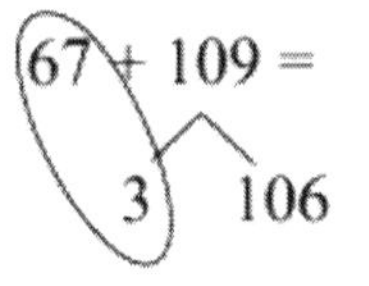

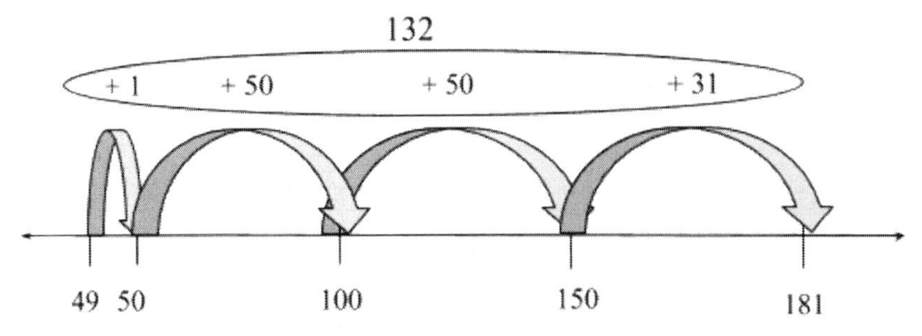

Figure 5.41.

5) Partial Sums (see figure 5.40)

Subtraction: 181 – 49

1) **Relate to Addition** (see figure 5.41)
2) **Decomposition** (see figure 5.42)
3) **Overestimation** (see figure 5.43)
4) **Constant Difference** (see figure 5.44)
5) **Partial Differences** (see figure 5.45)

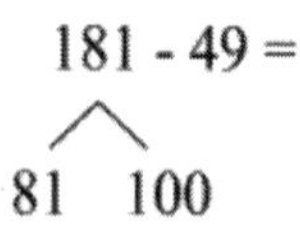

First, the child broke apart the minuend to help them subtract 'friendly' numbers.

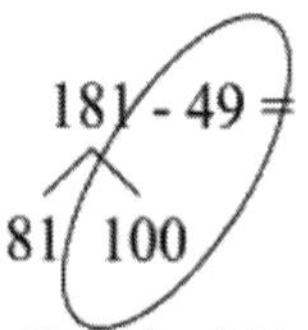

Next, the child subtracted 49 from 100, since they can do that mentally.

$$100 - 49 = 51$$
$$51 + 81 = 132$$

Last, the child added the remaining 81 to the difference of 100 and 49.

Figure 5.42.

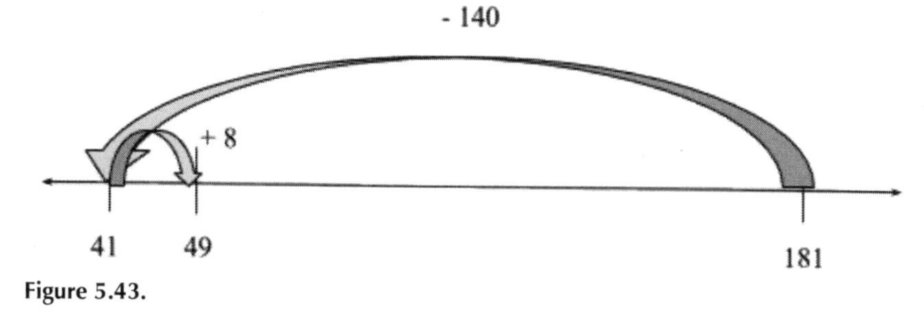

Figure 5.43.

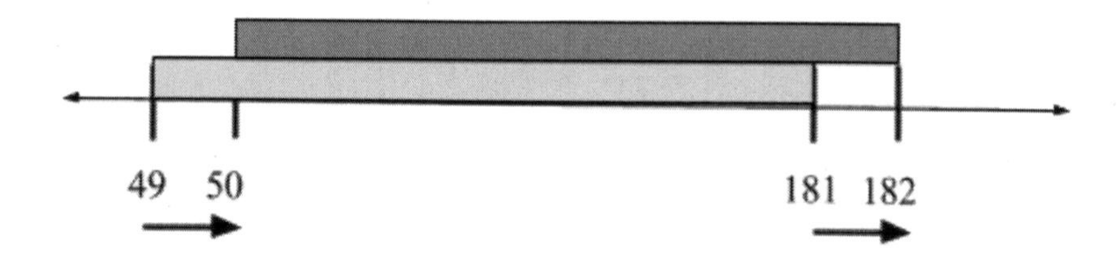

By adding 1 to each quantity in the expression, all we are doing is shifting the distance, not altering it.

Figure 5.44.

$$\begin{array}{rcl} 181 & = & 100 + 80 + 1 \\ \underline{-\ 49} & = & \underline{ - 40 - 9} \\ & & 100 + 40 - 8 \Rightarrow 132 \end{array}$$

Figure 5.45.

Chapter 6

Understanding Whole Number Multiplication and Division

"Children must be taught how to think, not what to think."

—Margaret Mead

Just like with addition and subtraction, we have encountered many frustrated parents who just want to see something familiar. We empathize with you and want you to be able to communicate with your child around the mathematics they are learning. Similar to chapter 5 (whole number addition and subtraction), we will illustrate a variety of strategies students might learn to multiply and divide whole numbers. But, before we do, we have to clarify why educators are pushing for conceptual understanding before learning the traditional U.S. Standard Algorithms. When we say standard algorithms in regard to multiplication or division, we are referring to the following, the way most American parents of today learned how to multiply and divide when they were in school (see figure 6.1).

Again, do not panic! Your child *will* still learn these methods, but not until later in their schooling than when you did. Most students will learn the traditional multiplication algorithm in fifth grade and long division in sixth grade. We encourage you to trust educators everywhere and not teach your child the way you learned it until they are learning it that way in school. These standard algorithms tend to focus on answer-getting, and once students are answer-focused, they often do not want to learn for understanding.

Children today will enter a workforce that relies on complex problem solving, creativity, and critical thinking, not on the ability to be quick or fast at computing numbers. Because of this, we must adjust the way we teach our youngsters how to think about operating with numbers so they can be most successful in the future. We predict that, by developing a stronger number sense with multiplying and dividing whole numbers yourself, you will find

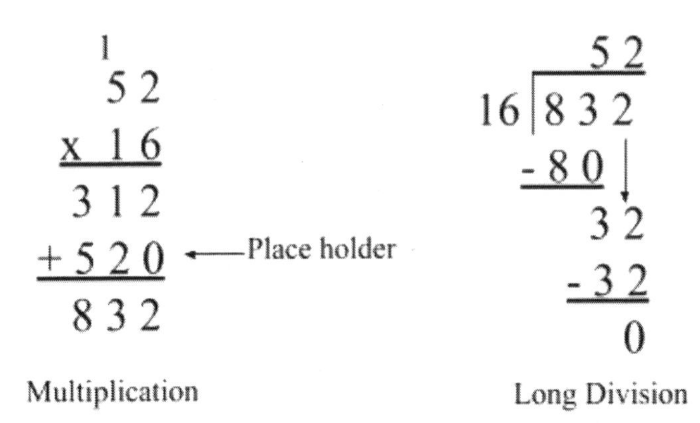

Figure 6.1.

that the procedures you once relied upon make more sense and that you will find yourself thinking flexibly in a real-world situation requiring you to multiply or divide, such as at the supermarket.

Did You Know?

There is a direct correlation between elementary school students' knowledge of fractions and whole-number division and their later mathematics achievement in high school. These correlations have been proven to be stronger indicators of future mathematics achievement than IQ, family education, and family income (Siegler et al., 2012).

THE OPERATIONS OF MULTIPLICATION AND DIVISION

To start, just like we did with addition and subtraction, we want you to take a moment to think deeply about the following questions. You may even want to write down your thoughts so you can refer back to your original thinking.

1. What is multiplication?
2. What is division?
3. Does multiplication always make things bigger?
4. Does division always make things smaller?

WHAT IS MULTIPLICATION?

So, how did you answer the first question? How would you define the operation of multiplication in mathematics? Not as easy to define as addition, is it? Many adults have a hard time explaining what multiplication is for two reasons: (1) we were just taught procedures and don't truly understand all the intricacies of the operation, and (2) multiplication is complex and has many different manifestations. Often, we hear adults and children say that multiplication is repeated addition, but only in some contexts is that accurate. When educators teach students something mathematical, one of our intents is to ensure it is *always* mathematically accurate. Learning that multiplication is repeated addition only works in certain instances and does not make much sense in others.

To illustrate our point, we invite you to think about the equation $4 \times 2 = 8$. In this case, we could create a compelling story that there were four baskets that each held two apples, resulting in eight apples altogether. Some might show this as $2 + 2 + 2 + 2 = 8$. Here, repeated addition certainly works. However, what about this equation: $\frac{1}{4} \times \frac{1}{2} = \frac{1}{8}$? Can you construct a compelling story that showcases repeated addition? Tricky, huh? That's because multiplication is multifaceted and challenging to define in one way.

Figure 6.2 shows three commonly described ways a person might interpret multiplication.

Multiplication is represented mathematically by the symbol "×" starting in grade 3, and then students learn in grade 6 to replace the "×" in some cases with "•" to avoid confusion with **variables**, or symbols that replace numbers. Additionally, students learn in sixth grade that multiplication is also implied when a number and variable are next to each other without an operation symbol. For example, $4y$ can be interpreted as 4 *times the quantity of* y, where y represents any real number, or as repeated addition of $y + y + y + y$, four times. Not until middle school (grade 6 and above) do students have to interpret a number or variable next to parentheses as representing multiplication. For instance, $2(x + 5)$ is the quantity of $x + 5$, doubled.

For the purpose of this chapter, we will use "•" to represent multiplication and focus solely on multiplication at the elementary level. Because students in grades 2 and below do not learn multiplication, this chapter will focus on students in grades 3 to 5 but will show how foundational understandings beginning in grade 2 are fundamental to solidifying a deep understanding of multiplication and division.

Starting in grade 3, students learn the mathematical language of multiplication. In most schools, students learn that a **factor** is a number that can be multiplied by another number to achieve a **product**, or answer, to a multiplication

Interpretation	Description	Visual
Multiplication as Repeated addition	When using multiplication to count quantities, one might repeatedly add quantities of equal size. For example, rather than counting 8 dots by ones, we might notice the organization of the dots and use the structure to form **equal groups** which will help us **skip count** to find the total dots quickly.	or Students can imagine 8 dots and organize them into 4 groups of 2 or 2 groups of 4 to skip count to find the total number of dots. Students might also build an **array** to show 2 equal groups of 4 dots or 4 equal groups of 2 dots. Both groupings show 8 dots in all.
Multiplication as Area	When interpreting multiplication as unknown area, one might be thinking geometrically about a rectangle. To find the area of a rectangle, we multiply the length by the width. In this thinking, students are identifying the amount of space within a rectangle.	4 ft 2 ft 8 ft^2
Multiplication as Scaling	Perhaps when you are multiplying, you are thinking about a multiplicative comparison (e.g., an object or a feature of an object stretching or shrinking). This is thinking about multiplication as **scaling**. To scale something is to **enlarge** it or **reduce** it. When thinking about scaling, students will use the phrases "times as many, times as much as, or times the size of" to describe the situation. In this example, Line B is 4 times the size of Line A.	8 in 2 in A B

Figure 6.2.

fact. Less commonly heard in schools is **multiplier** and **multiplicand**, which also represent the factors. We recommend you use the terms your child is learning at school so you can model precise language (see figure 6.3).

As we saw with addition and subtraction, one of the expectations of math students today is the ability to communicate precisely. One way we encourage students to do this is by using the math words assigned to the numbers that represent them. This helps students make more sense of the work they are doing and helps others interpret their thinking.

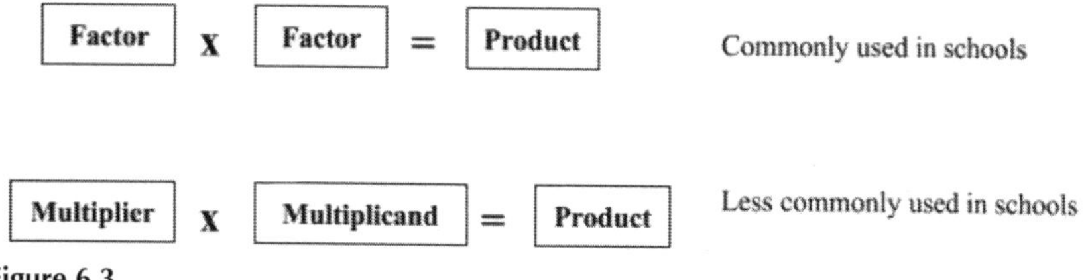

Figure 6.3.

TIP

It may assist you to refresh your memory of the five properties of numbers that involve multiplication as you continue to read this chapter. Students in grade 3 and above should be able to use these properties of numbers to think flexibly and problem solve. The properties are the commutative, associative, multiplicative identity, distributive, and the zero properties.

The *commutative property of multiplication* tells us that the order of the factors does not matter in finding the product. For example, 3 • 4 = 4 • 3, where both sides equal 12. Sometimes students in grade 3 call this the "flip-flop" property. It's okay for kids to use this terminology, but remember, our goal is to communicate precisely in math, so if your child says that, respond by saying, "That's right! That's what we call the commutative property of multiplication because the order doesn't matter!" It is expected by grade 5 that students know the mathematical term.

The *associative property of multiplication* tells us that the product is the same regardless how the factors are grouped. For instance, (2 • 3) • 1 = 2 • (3 • 1). Notice how we did *not* change the order of the numerals. All we did was regroup the numbers we were multiplying. By grouping the parentheses differently, we are really saying 6 • 1 = 2 • 3, which both equal 6.

The *multiplicative identity property* tells us that any time we multiply one by a quantity, the product is always the factor other than 1. For example, 4 • 1 = 4. When you hear the word *identity* think of it defining who a person is. In this case, it's defining what the product is (its identity remains the same).

The *distributive property* tells us that we can multiply a sum or difference by multiplying each part separately and then adding the products. For an example, see figure 6.4.

The distributive property of multiplication is a very important and frequently used property that is critical for algebraic thinking. Pay close

$2 \bullet (5 + 3) = 2 \bullet 5 + 2 \bullet 3 = 10 + 6 = 16$ **Multiplying a sum**

$2 \bullet (5 - 3) = 2 \bullet 5 - (2 \bullet 3) = 10 - 6 = 4$ **Multiplying a difference**

Figure 6.4.

attention throughout this chapter to see how the distributive property plays a key role in students' mental math with multiplication.

The *zero property of multiplication* tells us that any time we multiply a factor by zero, the product will be zero. For example, $5 \bullet 0 = 0$. Kids love learning this property because it means they don't need to know as many multiplication facts!

Now that you have a deeper understanding of the operation of multiplication, think back to the third question we asked at the beginning of the chapter: When multiplying, is the product always greater than its factors? Many children are taught as they begin learning multiplication that multiplying makes things bigger.

In third and fourth grade, this seems to be true, as they only work with **whole numbers**, or numbers not written as fractions (e.g., 0, 1, 2 . . .) and do not include negative numbers. However, this misconception expires as children enter fifth grade and multiply whole numbers by fractions less than one (*e.g.*, $4 \times \frac{1}{2} = 2$) or fractions less than one by fractions less than one (*e.g.*, $\frac{1}{3} \times \frac{1}{2} = \frac{1}{6}$).

This concept also fails as students enter sixth grade and begin multiplying with **integers**, or positive and negative numbers not in fraction form (e.g., $4 \times (-2) = -8$). In all of these examples, multiplication did not result in a larger product. Therefore, we encourage you to ask your child, "What's the context?" to help them make sense of the problem. If your child is solving a purely computational problem (just multiplying numbers without context) then ask your child to create a story that would help them make sense of the situation.

WHAT IS DIVISION?

The inverse operation of multiplication is division. This means that it is the opposite of multiplication. Just like multiplication, division is multifaceted and can be interpreted in many ways depending upon the context of the problem. Look at figure 6.5 to see a description and a visual of three common interpretations of division using the division problem $12 \div 4$.

Interpretation	Description	Visual
Division as sharing	You might hear your child talk about 'fair sharing.' This type of division, more formally known as *partitive division*, uses the context of dividing, or **partitioning**, something into equal-sized parts. For example, if there are 12 apples and the apples are shared equally among 4 people, each person can get 3 apples.	There are 12 apples and 4 people. If we share the apples among the 4 people equally, then how many apples would each person get? 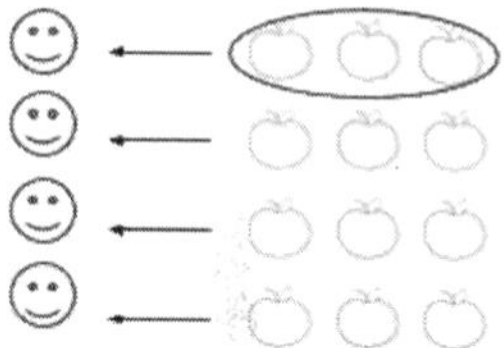 4 people 12 apples Each person would get 3 apples.
Division as repeated subtraction	Division as repeated subtraction is different than division as sharing because we do not know the number of groups for which we would share. In this case, we might subtract the same amount multiple times until we've identified the number of groups, or we might ask ourselves, "how many groups of __ fit into __?"	Bags can hold 4 apples each. If there are 12 apples, how many bags will be needed? 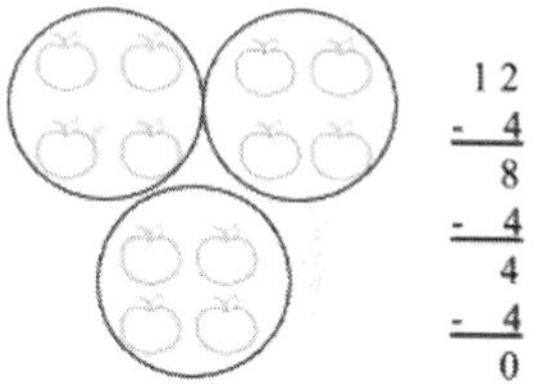 12 - 4 8 - 4 4 - 4 0 4 apples per bag Each time you make 1 bag, subtract 4 from how many are remaining 3 bags are needed because I repeatedly subtracted 4 apples from the total three times.
Division as Area	When interpreting division as area, one might be thinking geometrically about a rectangle. To find the missing side length of a rectangle, we divide the area by the width.	A rectangle has an area of 12 square feet and a width of 4 ft. What is the length of the rectangle? 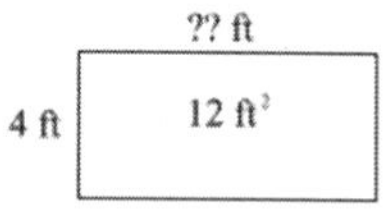 ?? ft 4 ft $12\ ft^2$ Area ÷ width = length

Figure 6.5.

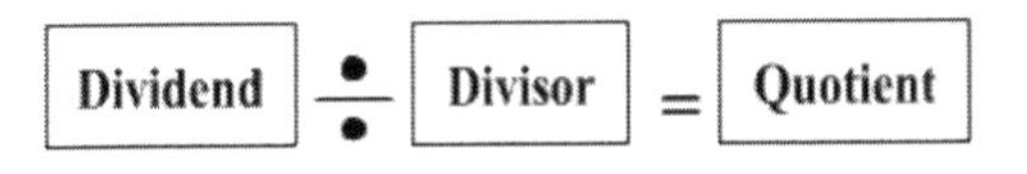

Figure 6.6.

Paying attention to the context of the problem might shift how students think about structuring their work and ultimately solving a division problem (e.g., *Am I given the number of groups? Do I know how many in each group? Is it easier to relate this problem to multiplication?*). It also helps students make sense of their own thinking to see if answers they get are reasonable given the context.

Just like with multiplication, we encourage teachers and parents to be precise with their mathematical language. Similar to the parts of a multiplication equation, in math, we have specific parts of a division equation (see figure 6.6).

The **dividend** is the number that will be divided. The **divisor** is the number being divided by the dividend. The **quotient** is the result of dividing the dividend and the divisor. Division is represented mathematically in the elementary grades by the symbol "÷," unless converting fractions to decimals, where students may interpret the line separating the numerator from the denominator (called a **vinculum**) as division.

Did You Know?

The symbol we generally associate with division has a name. This symbol (÷) goes by the mathematical name obelus (pronounced ob-uh-luhs). Teach your child the mathematically precise name of the symbol, and we guarantee they will impress their classmates!

Unlike multiplication, division is not commutative—this means that changing the order *does* affect the quotient. For example, $6 \div 3 \neq 3 \div 6$ because $2 \neq \frac{1}{2}$. Also, division is not associative—meaning that one cannot regroup the quantities and get the same result. For example, $(6 \div 3) \div 2 \neq 6 \div (3 \div 2)$ because $1 \neq 4$. Children begin to realize these properties through repeated exposure to story contexts in which they notice the math in a real-life situation.

The *zero property of division* tells us that zero divided by any real number, except itself, is zero. For example, $0 \div 2 = 0$.

The *identity property of division* tells us that any number divided by 1 is the number being divided. For example, $72 \div 1 = 72$.

Similar to multiplication, many children are initially taught division makes things smaller. Using arguments similar to the ones we used for multiplication, addition, and subtraction, it is critical that both teachers and parents explain concepts in ways that don't compromise future applications.

For example, in third and fourth grade, kids will only experience dividing whole numbers. Remember, whole numbers are numbers not written as fractions (e.g., 0, 1, 2 . . .) and do not include negative numbers. In third and fourth grade, students will only be exposed to situations where dividing does make things smaller.

But what happens when kids move on to fifth grade and divide a whole number by a fraction less than one? The result is not less! Furthermore, what happens when kids move on to seventh grade and they start operating with integers, or positive and negative whole numbers? The result won't always be less. For example, $-8 \div 4 = -2$, in which case you can see that -2 is not less than -8.

THE PROGRESSION OF MULTIPLICATION AND DIVISION

As with many areas of learning mathematics, developing a deep understanding of multiplication and division is best when students learn through a sequence, building on foundational knowledge and drawing connections to new information. Most schools today will follow a generally similar sequence as the one we describe in this chapter.

Please note that we label the progression by grade level—we do this because most curriculum standards today follow that same trajectory based on research of cognitive and developmental abilities. We advise you to ask your child's teacher at the beginning of the year what your child will learn regarding multiplication and division and use this chapter as a tool to support their learning. After we introduce the content and popular representations your child might learn, we will offer advice on what you can do at home to help your child make sense of multiplication and division.

FOUNDATIONAL UNDERSTANDINGS

In grade 2, most students will be exposed to **arrays**, or arrangements of objects in rows and columns, as a means of organizing a set of objects and also helping students move from counting by ones to **skip counting**, or counting forward or backward by a number other than 1. This skill helps children realize that they can *repeatedly add* the same number to get to a sum faster than counting by ones. Entering third grade with this understanding and ability to count by twos, threes, fours, fives, etc., will ease a child's ability to learn their multiplication facts.

To assist with this development, encourage your child to use counters to build or draw arrays. Look at figure 6.7 for an example.

Arrays are a great way to help your child start to make connections between addition and multiplication. Look back at figure 6.7 and think about how you see the total number of dots. Most likely, as an adult, you saw 24 dots because $3 \times 8 = 24$. As a beginning learner of multiplication or division, this might prove to be a challenging problem. This is a great time to build connections to the distributive property of multiplication to help use some of those other multiplication facts that your child might know from memory already. Look at figure 6.8 to see how by using two different colors or shades, you could make the task of knowing 3×8 much simpler for your child. As you look at figure 6.8, jot down what you notice.

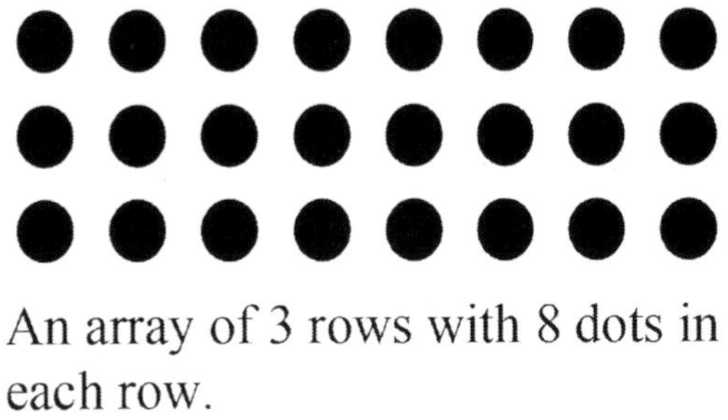

An array of 3 rows with 8 dots in each row.

Figure 6.7.

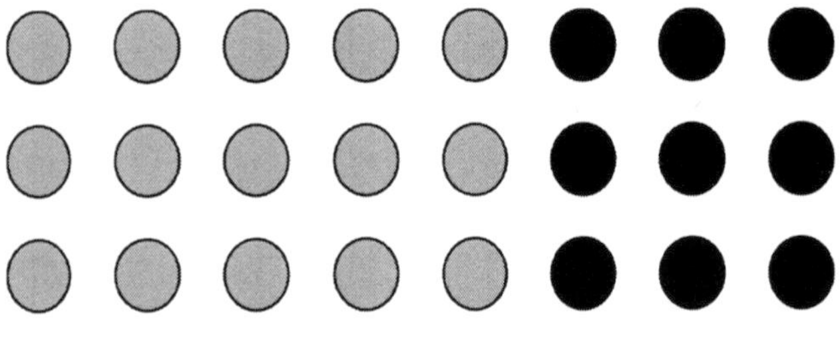

Figure 6.8.

Many students find counting by fives easier than most other counting sequences, so encourage your child to use their knowledge of the five sequence to help them solve new, unfamiliar problems, rather than using flashcards or memorizing facts in a rote manner. In the case of figure 6.8, students might notice that 3 groups of 8 is the same as 3 groups of 5 (gray dots) + 3 groups of 3 (black dots). This could be helpful if they know their count by 5s (3 groups of 5 is 15) and know that 3 groups of 3 is 9. Together, $15 + 9 = 24$, which is the same as 3 groups of 8.

Helping your child become familiar with arrays is a great starting place for multiplicative thinking, as the larger the array, the less and less kids want to individually count the dots (too tiresome!), forcing them to use their skip counting, and eventually connecting repeated addition to multiplication.

If your child is in second grade, start by using counters (see chapter 4) and strategically laying them out in rows and columns up to 5×5 (but remember, we aren't yet saying multiplication terms for second graders, so be careful of your language). If your child is in third grade or above but needs support with their basic facts (that's okay!), you can also use this strategy, but you *should* help them draw the connection of repeated groups to multiplication by writing the multiplication sentences that match and using multiplicative language. Figure 6.9 shows an example of how as a parent you can support student understanding of arrays.

If your child counts by ones (e.g., 1, 2, 3, 4, 5 . . .) let them do it, and then say, "Can you count by threes or fours for this?" For an added challenge, cover some of the counters with paper or a screen to encourage your child to visualize the array in their brain. Figure 6.10 shows an example.

We suggest you spend some time, even if your child is beyond second grade, returning back to skip counting and practicing counting forward and backward before honing in on memorizing facts. Have your child make up songs that help them remember the sequences, such as the following (to the tune of "Row, Row, Row Your Boat"):

> Four, eight, twelve, six—teen,
> twenty, twenty-four
> twenty-eight, thirty-two, thirty—six
> and for—ty.

Also have your child keep track of their count on their fingers. Keeping track on their fingers will help them make the connection that 4 is the first number in the count, 8 is the second number in the count, and so forth, leading right into them developing the understanding that $4 \times 1 = 4$, $4 \times 2 = 8$, etc.

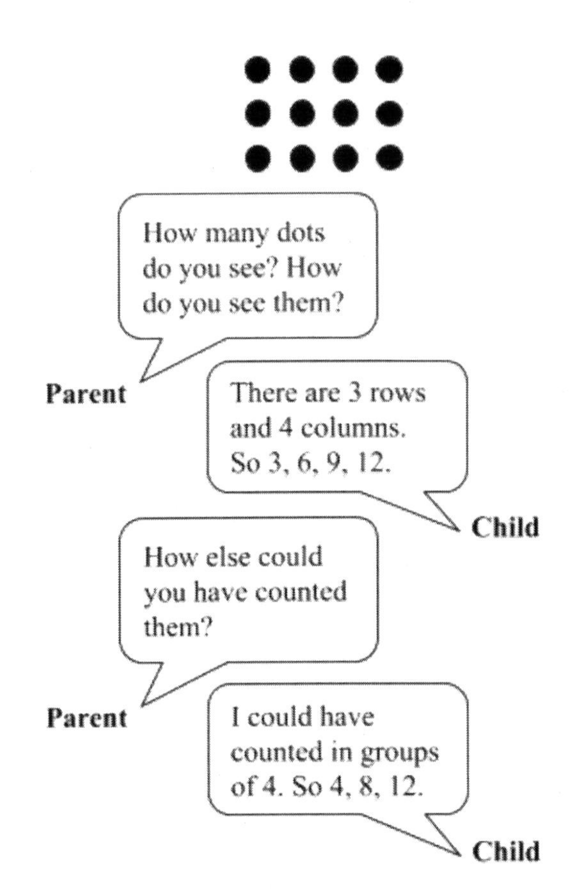

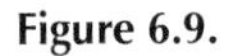

Figure 6.9.

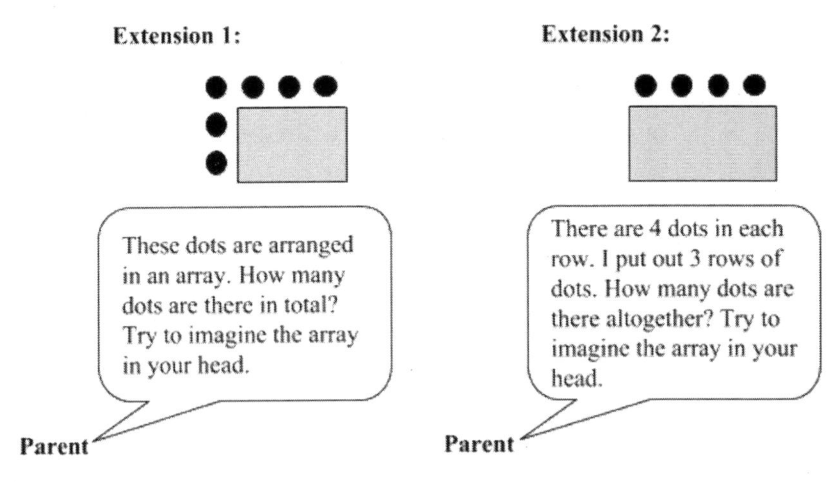

Figure 6.10.

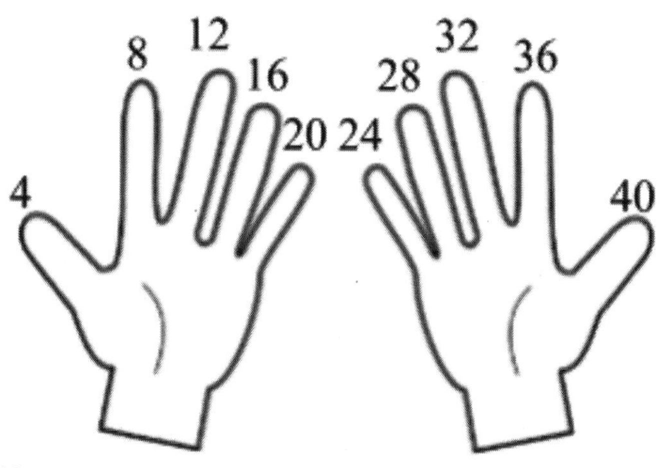

Figure 6.11.

As they sing, encourage them to keep track on their fingers (see figure 6.11).

You also want to help your child focus on the structure of the count. So, for example, if counting by fours, have them look at the sequence of numbers and tell you what they notice. More often than not, children are quick to notice a pattern, specifically the organization of the ones place (**4**, **8**, 1**2**, 1**6**, 2**0**, 2**4**, 2**8**, 3**2**, 3**6**, 4**0**, 4**4**, 4**8**, 5**2**, 5**6**, 6**0** . . .). If they don't notice a pattern, then leave it alone and come back to the same question later on. Children will make use of this structure as they develop a more conceptual understanding of multiplication and division in third grade. For example, by noticing this structure for counting by fours, a child may later be able to make the connection that 91 is not divisible by 4, one of the justifications being that it wouldn't fall in the count by four sequence if there's a 1 in the ones place. Eventually a child will come to realize that when counting by fours, all the numbers are even, giving them further validation that 91 is not divisible by 4 because it cannot be divided by 2, twice.

All this foundational support is vital for a child to learn with understanding. We caution you to not rush the process by having your child memorize their basic facts as you may compromise their understanding, making it challenging for them to later apply their knowledge to new contexts.

THIRD GRADE

Once students recognize that skip counting allows them to count faster than counting by ones, then they are ready to connect the skip counting sequence to multiplication as repeated addition of equal groups. Table 6.1 shows a

Table 6.1. Counting Sequence Structure and Relationship to Multiplication

Skip Counting	*4*	*8*	*12*	*16*	*20*	*24*	*28*	*32*	*36*	*40*
Multiplication Fact	4×1	4×2	4×3	4×4	4×5	4×6	4×7	4×8	4×9	4×10

visual representation of the structure of the counting sequence and its relationship to multiplication.

Do you see the connection between the skip counting sequence and multiplication facts for four? The first number in the sequence is 4, which is also 4 × 1. The second number in the sequence is 8, which is also 4 × 2 . . . etc. As children keep practicing their skip counting, many of them keep track of the count on their fingers and associate, for example, the fourth count with 16. This is helpful because when it's time to connect it to multiplication, they've essentially already been doing it! Now they just need to make meaning of "4 × 4 = 16" in an abstract sense. Children will be most successful in knowing their multiplication and division facts if they start early in second grade skip counting forward (repeated addition) and backward (repeated subtraction).

One-Digit by One-Digit Multiplication

By the end of third grade, students should know all their one-digit by one-digit multiplication facts and the related division facts. See figure 6.12 for all the one-digit by one-digit multiplication facts. *Note: Students will typically see this table flipped, with the greatest products in the bottom right corner. We believe students should read this table just like they would read a graph, so we present it structurally this way.*

We realize figure 6.12 looks daunting—seems like a lot a child must know! However, we ask you (and your youngster) to apply a growth mindset when thinking about the facts children need to know. Show figure 6.12 to your child and ask them what they notice about the structure of the table. One piece to notice is that the table is symmetrical. The boxes above the shaded diagonal have the same product as the boxes below the shaded diagonal. Children should connect this to the commutative property of multiplication and become excited because that means fewer facts that will need to be remembered.

Another part a child might notice is that any time we multiply a number by zero, the result is zero. Rather than tell your child that rule, why not have them notice that on their own? This structure is important because it means they don't need to remember any facts that have "times zero" because they have generalized a rule that always works!

Similarly, a child might also notice that any time we multiply a number by one, the result is the number we multiplied. Again, let your child come to that

9	0 x 9 = 0	1 x 9 = 9	2 x 9 = 18	3 x 9 = 27	4 x 9 = 36	5 x 9 = 45	6 x 9 = 54	7 x 9 = 63	8 x 9 = 72	9 x 9 = 81
8	0 x 8 = 0	1 x 8 = 8	2 x 8 = 16	3 x 8 = 24	4 x 8 = 32	5 x 8 = 40	6 x 8 = 48	7 x 8 = 56	8 x 8 = 64	9 x 8 = 72
7	0 x 7 = 0	1 x 7 = 7	2 x 7 = 14	3 x 7 = 21	4 x 7 = 28	5 x 7 = 35	6 x 7 = 42	7 x 7 = 49	8 x 7 = 56	9 x 7 = 63
6	0 x 6 = 0	1 x 6 = 6	2 x 6 = 12	3 x 6 = 18	4 x 6 = 24	5 x 6 = 30	6 x 6 = 36	7 x 6 = 42	8 x 6 = 48	9 x 6 = 54
5	0 x 5 = 0	1 x 5 = 5	2 x 5 = 10	3 x 5 = 15	4 x 5 = 20	5 x 5 = 25	6 x 5 = 30	7 x 5 = 35	8 x 5 = 40	9 x 5 = 45
4	0 x 4 = 0	1 x 4 = 4	2 x 4 = 8	3 x 4 = 12	4 x 4 = 16	5 x 4 = 20	6 x 4 = 24	7 x 4 = 28	8 x 4 = 32	9 x 4 = 36
3	0 x 3 = 0	1 x 3 = 3	2 x 3 = 6	3 x 3 = 9	4 x 3 = 12	5 x 3 = 15	6 x 3 = 18	7 x 3 = 21	8 x 3 = 24	9 x 3 = 27
2	0 x 2 = 0	1 x 2 = 2	2 x 2 = 4	3 x 2 = 6	4 x 2 = 8	5 x 2 = 10	6 x 2 = 12	7 x 2 = 14	8 x 2 = 16	9 x 2 = 18
1	0 x 1 = 0	1 x 1 = 1	2 x 1 = 2	3 x 1 = 3	4 x 1 = 4	5 x 1 = 5	6 x 1 = 6	7 x 1 = 7	8 x 1 = 8	9 x 1 = 9
0	0 x 0 = 0	1 x 0 = 0	2 x 0 = 0	3 x 0 = 0	4 x 0 = 0	5 x 0 = 5	6 x 0 = 0	7 x 0 = 0	8 x 0 = 8	9 x 0 = 9
	0	1	2	3	4	5	6	7	8	9

Figure 6.12.

conclusion on their own. When we discover or figure things out for ourselves, we tend to hold on to that knowledge. Why is this finding important? Well, if they know that rule, then again, your child doesn't need to spend countless hours memorizing the facts through flashcards! They simply need to recognize the rule they've developed.

There are many other noticings one could make with the multiplication table, so we encourage you to play a game with your child where you invite them to jot down all the patterns they see. If you do a quick search on the internet, you will find other patterns you could discuss with your child.

Two-Digit Dividends by One-Digit Divisors

Just like students should know their one-digit by one-digit multiplication facts by the end of third grade, students should also know the related division facts of two-digit dividends by one-digit divisors. Figure 6.13 shows a division version of the multiplication table. Most schools do not use division tables, but we find students work well with consistency and structure. As such, it's just as easy to create a division table for children to analyze. Have your child create their own, where each row must equal the quotient on the left.

Similar to the multiplication table, while this looks daunting, remember the structure of the table enables us to notice repetition, and therefore, kids who use the structure of the number system will be able to learn their facts more quickly because they will connect numerical relationships, such as the fact that dividing by 1 always results in the same number as the dividend.

What other structures in figure 6.13 do you notice that could help you reduce the number of facts one must remember?

Quotients									
9	9÷1 = 9	18÷2 = 9	27÷3 = 9	36÷4 = 9	45÷5 = 9	54÷6 = 9	63÷7 = 9	72÷8 = 9	81÷9 = 9
8	8÷1 = 8	16÷2 = 8	24÷3 = 8	32÷4 = 8	40÷5 = 8	48÷6 = 8	56÷7 = 8	64÷8 = 8	72÷9 = 8
7	7÷1 = 7	14÷2 = 7	21÷3 = 7	28÷4 = 7	35÷5 = 7	42÷6 = 7	49÷7 = 7	56÷8 = 7	63÷9 = 7
6	6÷1 = 6	12÷2 = 6	18÷3 = 6	24÷4 = 6	30÷5 = 6	36÷6 = 6	42÷7 = 6	48÷8 = 6	54÷9 = 6
5	5÷1 = 5	10÷2 = 5	15÷3 = 5	20÷4 = 5	25÷5 = 5	30÷6 = 5	35÷7 = 5	40÷8 = 5	45÷9 = 5
4	4÷1 = 4	8÷2 = 4	12÷3 = 4	16÷4 = 4	20÷5 = 4	24÷6 = 4	28÷7 = 4	32÷8 = 4	36÷9 = 4
3	3÷1 = 3	6÷2 = 3	9÷3 = 3	12÷4 = 3	15÷5 = 3	18÷6 = 3	21÷7 = 3	24÷8 = 3	27÷9 = 3
2	2÷1 = 2	4÷2 = 2	6÷3 = 2	8÷4 = 2	10÷5 = 2	12÷6 = 2	14÷7 = 2	16÷8 = 2	18÷9 = 2
1	1÷1 = 1	2÷2 = 1	3÷3 = 1	4÷4 = 1	5÷5 = 1	6÷6 = 1	7÷7 = 1	8÷8 = 1	9÷9 = 1
	1	2	3	4	5	6	7	8	9
	Divisors								

Figure 6.13.

Connecting Arrays to Area Models

As students begin to develop familiarity with their multiplication and division facts, they will also be drawing connections between their work in second grade with repeated addition and their new work in third grade with multiplication. As your child works through third grade, a critical focal point becomes recognizing area as an attribute of rectangles, which you will see will have a profound influence on their multiplication and division computation. Having a solid foundation of their basic facts (multiplication up to 10 × 10 and division of two-digit by one-digit) will help them make sense of future work.

Think back to the arrays students were using to build the connection between addition and multiplication. After students are exposed to arrays and have developed a stronger understanding of skip counting, they will begin to see connections to multiplication as area through the use of **area models**, or diagrams that represent the dimensions of a rectangle as factors and the area as the product, and ultimately, use the area model to identify the side length of the rectangle given the area and the width (division). Figure 6.14 shows the progression of arrays to area models.

Can you see how the array leads to the beginning area model? The array had countable objects and helped students keep track of groups when counting. Now, to build the beginning area model, students learn that, for a shape to have area, there cannot be gaps, spaces, or overlaps and that the area is

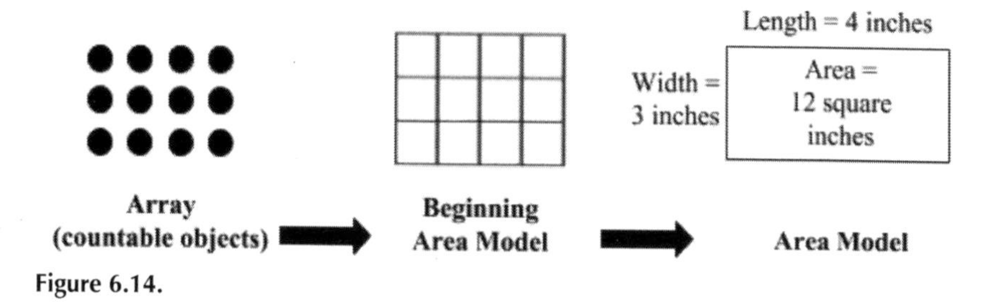

Figure 6.14.

dependent upon the number of equal-sized squares inside the rectangle. Finally, moving from *concrete* to *representational* to *abstract*, students just see a rectangle with numbers and recognize that the length multiplied by the width results in the area of a rectangle.

There's one more connection we want to share with you before we move deeper into multiplication and division. We particularly like the area model as a representational model for students to use because it helps them solidify their operations with area. We find lots of students mix up perimeter and area, but in the classrooms where we see students making strong connections between area models and multiplication and division, we find these mix-ups occur less frequently. This understanding all happens very gradually, so be sure to let your child's teacher lead the learning, and now you have the language to be able to support the understanding at home. We will come back to area models throughout this chapter.

One Digit by Multiples of 10

After students have begun to develop their basic multiplication facts through 10 × 10, they then start multiplying one-digit numbers by 10 and **multiples** of 10 (e.g., 20, 30, 40 . . .), with the basic idea focusing on place-value reasoning. In grade 3, students work within the range of 10–90, but in grade 4, students expand to all multiples of 10. Our number system is fascinating, and we need students to recognize the value *ten* has so they can use it to make sense of operating with numbers.

Look at figure 6.15 and think about what you notice. What is staying the same? What is changing?

Students who already know 4 × 1 can be very successful finding the value of 4 × 10, 4 × 100, and 4 × 1000 because they already have the basic structure and realize that each expression is *10 times* the amount before it! Now, when we were kids, we were taught to "move the decimal point" or "add a zero to the end of the number." We want to emphasize how important it is that you avoid using the language that promotes tricks. Do you see the decimal

	Thou.	Hund.	Tens	Ones	.
$4\times1000 =$	4	0	0	0	.
$4\times100 =$		4	0	0	.
$4\times10 =$			4	0	.
$4\times1 =$				4	.

Figure 6.15.

point in our chart move? No, because the decimal point doesn't really move; instead, as you can see from figure 6.15, the digits are shifting place values. Similarly, we didn't just "add a zero" because we know when we add zero to any number the value remains unchanged. Instead, we can see that each successive row is 10 times the size of the row before.

Helping children develop this reasoning will come in handy as they multiply larger quantities, use the metric system, and make use of the distributive and associative properties. One way you can help your child at home is by helping them see relationships between numbers. For example, if your child asks you for help solving 9 × 600, rather than focusing on the answer, we encourage you to ask, "Well, what is 9 × 6?" and then build onto that by asking, "So if 9 × 6 is 54, then what is 9 × 60? What about 9 × 600?"

You can also use the actual number words to help make the point clear by asking, "What is 9 groups of 6 ones? [54 ones]. So, what is 9 groups of 6 tens? [54 tens]. Okay, so what is 9 groups of 6 hundreds? [54 hundreds]. Can you rewrite 54 hundred as a number?" This idea builds on the *associative property of multiplication* (refer back to the tip box at the beginning of the chapter) because you are regrouping the expression 9 × 600 to really say (9 × 6) × 100.

By responding to your child with a question, and in particular one that hones in on understanding, not answers, you will help your child learn how to persevere through challenging problems and embrace a growth mindset.

Multiples of 10 Divided by One Digit

Similarly, students should be thinking about what happens when we divide multiples of ten by one-digit numbers. Let's look at the division version of the place-value chart shown in figure 6.15 so we can see a comparison. Looking at figure 6.16, again, we can see that the decimal does not move and that each row is $\frac{1}{10}$ (one-tenth) the size of the row before it.

	Thou.	Hund.	Tens	Ones	.
$4000 \div 4 =$	1	0	0	0	.
$400 \div 4 =$		1	0	0	.
$40 \div 4 =$			1	0	.
$4 \div 4 =$				1	.

Figure 6.16.

How can you use this information to help your child at home? Avoid tricks. Catch yourself as you start saying "move the decimal point" or "add/cross out a zero." Instead, help your child see the structure of the number system by asking them questions about the basic facts hidden within. For example, in solving $400 \div 4$, say, "You know $4 \div 4 = 1$," so now we need to think about the place values: $400 \div 4$ is 100 times as much as $4 \div 4$. Another way you could phrase your questioning is to ask, "How could you rephrase $400 \div 4$ to be a multiplication problem?" Since $400 \div 4$ is the same as asking 4 times what is 400, children can use their multiplicative thinking to answer a division problem.

FOURTH GRADE

Multiplying One Digit by Multiple Digits

Once students are nimble with their basic multiplication facts and have explored multiplication with both one-digit by one-digit and one-digit by multiples of 10, then they are ready to apply their previous knowledge to one-digit by multi-digit multiplication problems. Multi-digit, as you might be able to infer, are numbers with more than one numeral (e.g., 32, 453, 8674, etc.). The area model is a great representation to continue developing this understanding. *Note: Some teachers and curriculum call the area model, when drawn with rectangles, the "box method." This is the same as the area model that we refer to throughout this book.* Figure 6.17 shows one way to use the area model to find 8×748.

Using the area model to solve 8×748 allows a child to build on prior understanding and rewrite the problem as a sum of **partial products**, or the adding together of simpler multiplication problems. Look at figure 6.17 to see what we mean. When attempting to calculate 8×748, using an area model

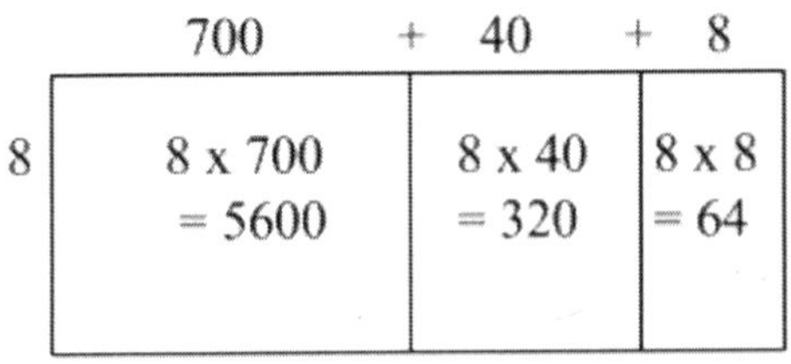

	700	+ 40	+ 8
8	8 x 700 = 5600	8 x 40 = 320	8 x 8 = 64

Add the areas of the three rectangles together to find the area of the entire rectangle.

$$5600 + 320 + 64 = 5984$$

Figure 6.17.

encourages students to think about the expression as 8 representing the width of a rectangle (inches, feet, whatever the unit) and the length as 748. Essentially, they are looking for the area, which they know is the space inside of the rectangle. To make this simpler, oftentimes children will decompose, or break apart, the factors into their place values, or other configurations that make the partial products easier to compute.

With 8 × 748, you can see that, in this case, the child might decompose 748 into 700 + 40 + 8. This then allows the student to think about 8 × 700, 8 × 40, and 8 × 8, instead of what could be daunting as 8 × 748. Notice how by breaking apart 748 by place values, the child, at minimum, needs to know their basic facts and have an understanding of place values to be successful. The area model also highlights a child's deep knowledge of the *distributive property*. This means that they understand that 8 × 748 is the same as multiplying 8 by smaller parts that total 748. If a child truly understands how to multiply a one-digit number by a multiple of 10, then they will be very successful using the area model to help them visualize the partial products.

Following through with 8 × 748, children who break apart 748 into its **place values** (7 hundreds, 4 tens, 8 ones) would perhaps start by asking themselves what 8 × 700 equals. They recognize that this is the area of part of the larger rectangle. They continue with 8 × 40 and then 8 × 8 and add all the areas together to identify the area of the whole rectangle.

Once students have had repeated exposure and practice to the area model (in school they will start with smaller numbers and actually build rectangles with blocks, then draw the pictures to match), students might be exposed to writing the partial products in an algorithmic way to help build a bridge toward the standard algorithms with which most adults are so comfortable.

Figure 6.18 shows how a child might be taught to move from using an area model to an algorithm to document the partial products.

Do you see the similarities between the area model and an algorithm for partial products? They are showing the same concept, but when we write algorithms, they are often devoid of the visuals that kids can find very helpful. For this reason, children usually become familiar with the area model and then start to record their steps abstractly in an algorithmic way. By writing the algorithm in this way, kids are forced to focus on the place values of each digit rather than just looking at the digits as numbers. This becomes powerful later on when children learn the standard algorithm and are able to make meaning of the steps and are able to articulate *why* each step works.

To help students begin to draw connections to the standard algorithm, students might also learn to write their partial products starting with the ones place.

Do you see the difference between the algorithm in figure 6.18 and the algorithm in figure 6.19?

Hopefully you noticed that the order of the partial products is reversed. Writing the partial products in this manner sets students up to begin to learn the standard algorithm and make sense of it. Notice how in writing the partial

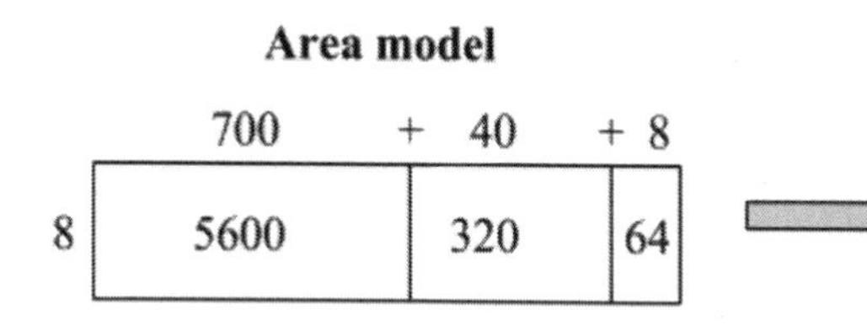

Algorithm

```
  748
x   8
 5600 = 8 x 700
+ 320 = 8 x 40
   64 = 8 x 8
 5984
```

Figure 6.18.

```
  748
x   8
   64 = 8 x 8
+ 320 = 8 x 40
 5600 = 8 x 700
 5984
```

Figure 6.19.

products this way, students start by multiplying the ones places, just as they would in the standard algorithm. There is not just one way to write the partial products because the order for addition doesn't matter, but we encourage you to follow the lead of your child's teacher so your child has consistency.

Dividing Multiple Digits by One Digit

Just like we have done throughout the chapter, let's now connect what you just saw with multiplication to division. The area model still serves as a useful visual representation, and again, keep in mind that, in school, many teachers will have students use base-ten blocks to actually build the rectangles so they can see the area model concretely and then represent it by drawing it. Let's use the problem 5984 ÷ 8 to think about division as area. When thinking about division as area, we are really asking ourselves, "If the area of the rectangle is 5984 and the width is 8, what is the length of the rectangle?" Figure 6.20 shows how the area model in this case looks so similar to the long division problem we are used to setting up as adults.

Continuing with the area model, the question becomes "How many hundreds can I multiply by 8 to get close to 5984?" Students at this point need to be fluent with their basic facts so they can estimate wisely. At this point, a student might be thinking, "Hmm . . . 8 × 600 = 4800, 8 × 700 = 5600, 8 × 800 = 6400 . . . it must be 8 × 700 because 8 × 800 results in more area than we have and 8 × 600 means there is more space to be uncovered." So, a student will partition the area model to show how many hundreds and what remains (see figure 6.21).

Thinking in the same terms as before, a student might ask, "How many tens can I multiply by 8 to get close to 384?" Using their knowledge of basic facts, they will see that 8 × 40 = 320, which is the closest they can get without going over the number of tens allotted. Figure 6.22 shows how a student might notate that thinking.

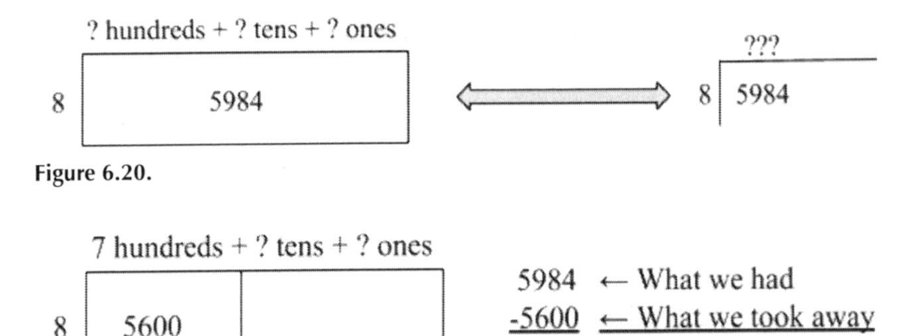

Figure 6.20.

Figure 6.21.

At this point, the child would recognize that $8 \times 8 = 64$ and fill in the remainder of the area model (see figure 6.23).

Take a look back at figure 6.18. Do you see that this is the exact same figure that we had for multiplication? All we did was work backward to figure out the length, instead of multiplying the length and the width to find the area.

The area model for division shows the **partial quotients** (or the subtraction of simpler division problems) by place value. Partial quotients can also be written in an algorithm, just like partial products. In fact, we've been writing the steps in each figure, but figure 6.24 shows that many students are learning how to write an algorithm.

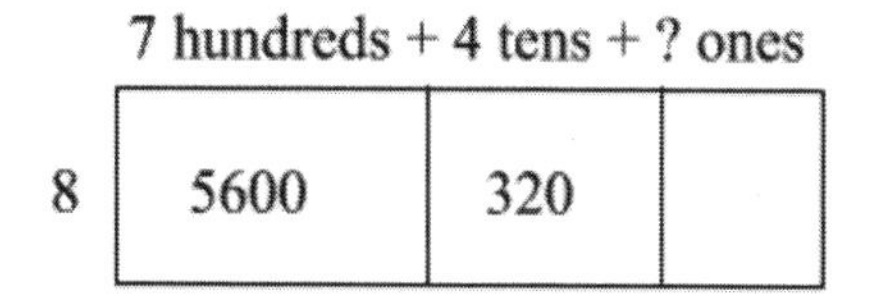

7 hundreds + 4 tens + ? ones
8
5600
320
5984 ← What we had
-5600 ← What we took away
384 ← What remained
- 320 ← What we took away
64 ← What remains

Figure 6.22.

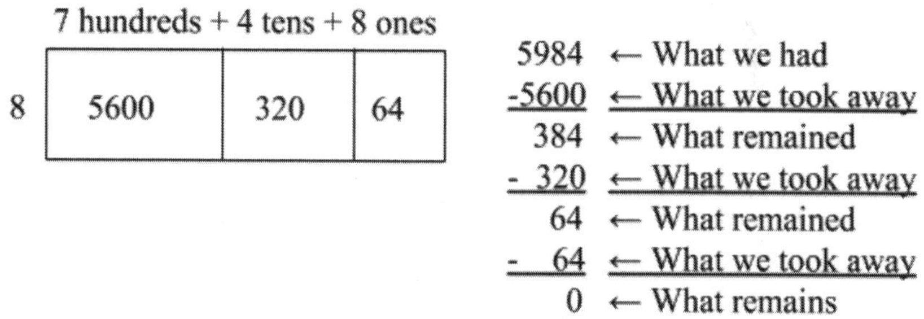

7 hundreds + 4 tens + 8 ones
8
5600
320
64
5984 ← What we had
-5600 ← What we took away
384 ← What remained
- 320 ← What we took away
64 ← What remained
- 64 ← What we took away
0 ← What remains

Figure 6.23.

Area Model
? hundreds + ? tens + ? ones
8
5984
Algorithm
???
8
5984
- 5600
384
- 320
64
- 64
0
8 x 700
8 x 320
8 x 8

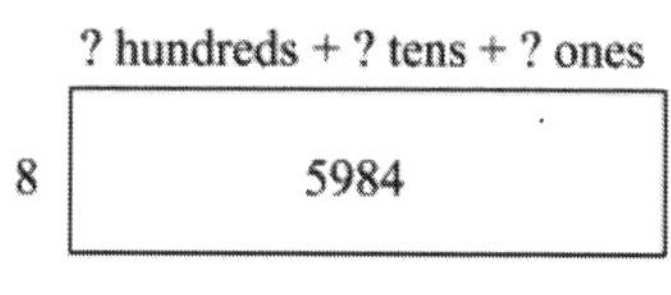

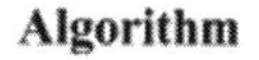

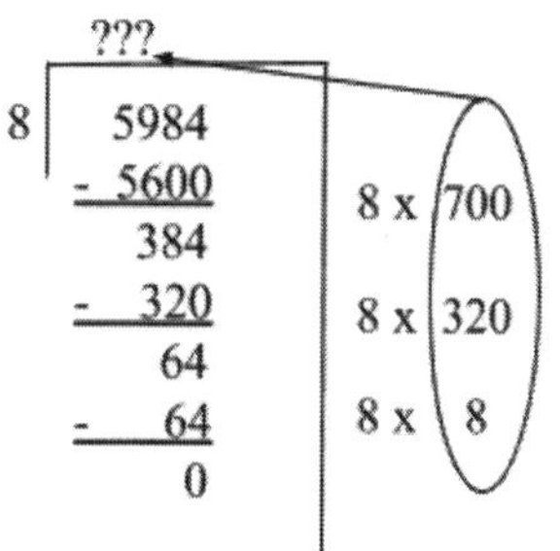

Figure 6.24.

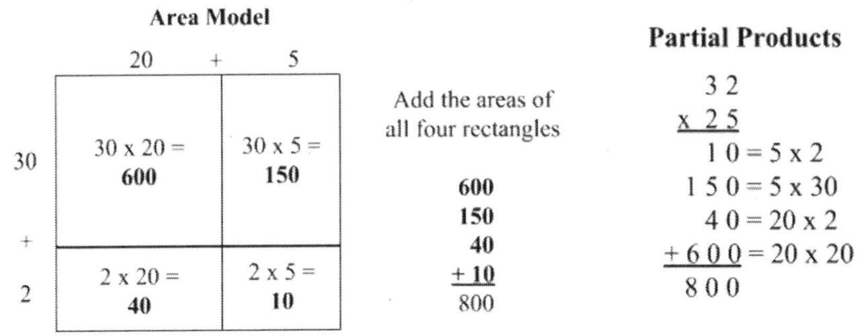

Area Model

	20	+ 5
30	30 x 20 = **600**	30 x 5 = **150**
+ 2	2 x 20 = **40**	2 x 5 = **10**

Add the areas of all four rectangles

600
150
40
+ 10
800

Partial Products

3 2
x 2 5
1 0 = 5 x 2
1 5 0 = 5 x 30
4 0 = 20 x 2
+ 6 0 0 = 20 x 20
8 0 0

Figure 6.25.

You'll note it almost looks like there's a large "7" or "hangman" post in the partial quotients algorithm setup—many teachers have found it helpful to take the division bracket and draw a long vertical line to help kids separate the partial quotients from the partial dividends. What is circled are the partial quotients that a student would add together to get the total quotient. If your child's teacher doesn't write the partial quotients in this manner, we suggest you ask the teacher for an example. The goal is eventually to see the connection to the long division algorithm, but for now, it's just an organized way to write down all the steps.

Multi-digit by Multi-digit Multiplication and Division

As children become more fluent with one-digit by multi-digit multiplication, they move into multi-digit by multi-digit multiplication and division, but the representations and models they use remain the same! That's the beauty of these tools—they offer longevity. When you get to chapter 8 (beyond elementary) you will see how the area model is used in both middle and high school.

To explore multi-digit by multi-digit multiplication, let's look closely at 32 × 25. Based on what you've seen so far, can you use the area model to solve? Look at figure 6.25 to see if you were right! Can you use partial products to solve? Check your work.

At this point, students are typically feeling comfortable enough with two main strategies that they can also start applying their number sense to problems. For example, when thinking about 32 × 25, children might notice that they take half of 32 and double 25 and keep the product the same. Figure 6.27 shows what we mean.

In the example shown in figure 6.27, a child noticed that they could cut the original rectangle horizontally in half to get two equal areas of 16 × 25.

Did You Know?

Ancient Egyptians used to multiply using a form of partial products. Their process required no memorization of multiplication facts, just pure understanding of doubling! Figure 6.26 shows how the Egyptians would have solved 21 × 18. (*Note: This is in modern notation.*)

21 x 18		
1	18	18
2	36	
4	72	72
8	144	
16	288	288
	21 x 18 =	378

Figure 6.26.

First, they would make two columns. In the left column, they'd start with 1 and keep doubling until they got closest to the multiplier (e.g., next would have been 32, which is greater than 21). In the right column, they'd double the multiplicand. Then, they find the sum of the multiplier (21) within the doubled numbers (in this case, 1 + 4 + 16 = 21) and add their corresponding values from the right column to find the product.

By rearranging the rectangle after they doubled and halved, they now have a 16 × 50 rectangle, which they know has the same area as the 32 × 25. To solve, 16 × 50 is a lot easier for kids at this point to compute mentally because they know how to multiply by multiples of 10. In this case, they either might just know 16 × 5 from memory, or they might count by fives

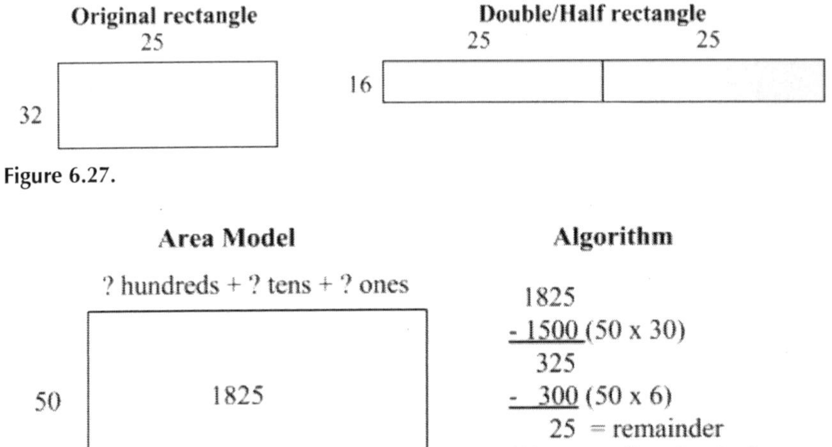

Figure 6.27.

Algorithm

1825
- 1500 (50 x 30)
325
- 300 (50 x 6)
25 = remainder

**You cannot have hundreds because 50 times anything in the hundreds would be too big!*

So . . . 36 (3 tens, 6 ones) is the answer with 25/50 or one-half remaining.

Figure 6.28.

to realize 16 × 5 = 80 and they know that the answer needs to be *10 times greater* because they were working with 50, not 5.

Now taking a look at multi-digit by multi-digit division, let's focus on the problem 1825 ÷ 50. Can you use what you've learned so far to demonstrate how to solve 1825 ÷ 50 with the area model and partial quotients? You might be recognizing that 1825 isn't **divisible** by 50, meaning that there is no one whole number that can be multiplied by 50 to get 1825. This means you will have a remainder. As you solve, consider what to do with the remainder. Go ahead, really challenge your brain! Use another piece of paper and refer back to earlier problems in the chapter to help you make sense of it. When you're ready, look at figure 6.28 to check your work.

How can you help your child at home with this idea? Be open minded to multiple paths to get the same answer. Be flexible about how your child computes their math, even if it seems at first inefficient or time consuming. Know there is a larger takeaway behind the strategy.

FIFTH GRADE

By the time your child gets to grade 5, their number sense and ability to multiply and divide numbers mentally should blow your mind! Children today are being forced more than ever to reason with numbers, and as a result, their mental math capabilities are often much quicker than adults'! The expectation in grade 5 is for students to be able to multiply using the standard algorithm by which most American adults are most familiar. It's critical that teachers build a nice bridge and connection to the standard algorithm so that students *understand* how it works, not just memorize a bunch of steps.

Check out figure 6.29 to see how we can bridge a child's understanding by allowing them to notice connections between the area model, an algorithm for partial products, and the standard algorithm for multiplication.

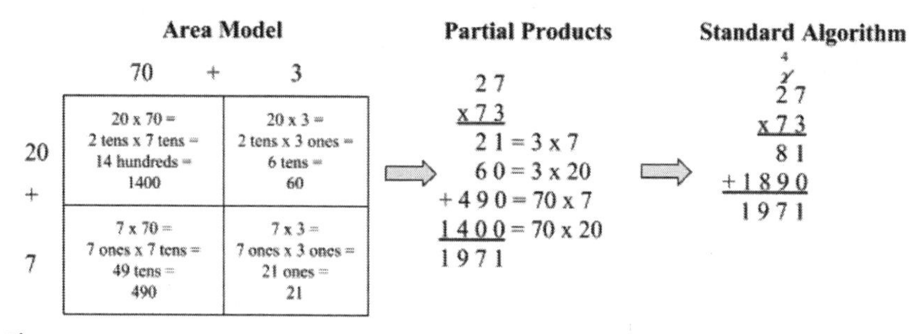

Figure 6.29.

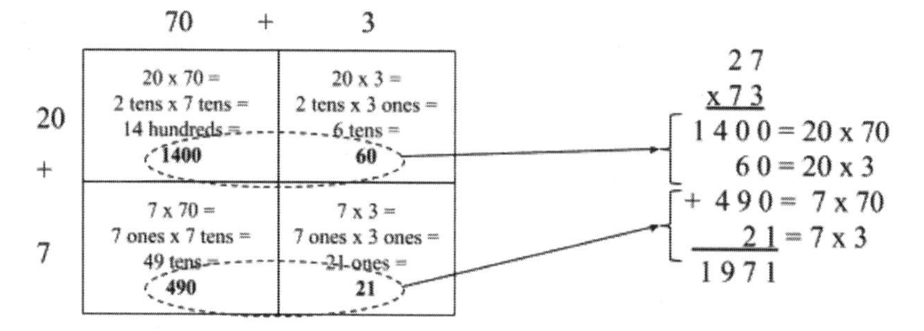

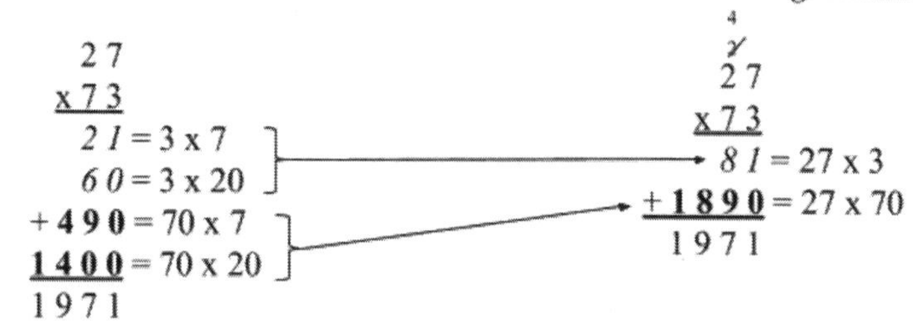

Figure 6.30.

Can you see the connection? What do you notice is the same about the three representations? What's different?

Next check out a closer version in figure 6.30, showing the transfer of knowledge from area models to partial products and then from partial products to the standard algorithm.

When it comes to division, grade 5 students are still using the area model and partial quotients to solve division problems. It is not until they enter grade 6 that they are expected to use the standard algorithm for division (long division). However, it is at this point that students will now start to think about what would happen if they multiply and divide with decimals and fractions, which we will showcase in chapter 7 (fractions).

Chapter 7

Fractions

> "The losses that occur because of the gaps in conceptual understanding about fractions, ratios, and related topics are incalculable. The consequences of doing, rather than understanding, directly or indirectly affect a person's attitudes toward mathematics, enjoyment and motivation in learning, course selection in mathematics and science, achievement, career flexibility, and even the ability to fully appreciate some of the simplest phenomena in everyday life."
>
> —Susan J. Lamon (2012, p. xi)

Research shows that students need exposure to different representations of numbers in order to develop what is commonly referred to as "number sense," or flexibility with numbers (Lesh, Landau, & Hamilton, 1983). Fractions also represent numbers, and so students need the same type of experiences as they had with whole numbers to be able to manipulate and use fractions with understanding.

Fractions are often a source of frustration during elementary school, and it is common to hear middle and high school teachers reflecting on how their students don't know how to operate with fractions. Many times, this is because students have previously been taught procedures and algorithms, and like much in life, if they don't use it, they lose it, or they confuse those procedures with others learned for whole numbers. This is exactly why building the understanding and foundation for number sense has been shown to increase student achievement later on.

As a parent who is looking to help their child with better understanding the math instruction of today, it is just as important that you develop a more sound understanding, too. Our intent is not to run through an entire fractions course with you in this chapter but, rather, to expose you to foundational understandings of fractions that you may see your child using, so that you are equipped to communicate more precisely with your child. To do this, we will first explore the meaning of fractions (What are they? How can we interpret them in a given problem?) and then compare fractions using conceptual understanding. We will then draw connections to the previous chapters by demonstrating various ways to operate with fractions.

THE MEANING OF FRACTIONS

In order for students to really understand fractions, it is critical that they learn to view fractions as numbers, and specifically, numbers that represent different constructs based on the context. As an adult, take a moment and think about the number $\frac{4}{5}$. Can you think of different contexts where that fraction represents different things? Jot down a few examples.

There are five common interpretations of fractions. Some you may be familiar with, but you most likely learned them as math topics in isolation and not in the context of representing fractions. Understanding these five ways fractions can be used will allow you to solidify the meaning of fractions and be better equipped to help your youngster.

Table 7.1 is meant to give you an idea of how complex fractions really are and to remind you that, even as adults, we are still learning more about fractions. Students do not need to be exposed to the table nor do they need to ever identify the interpretation of the fraction they are using. For example, students will not need to say, "in this case, $\frac{3}{4}$ is acting as an operator." But, as an adult, having an overview that reminds us that fractions can manifest themselves in many different contexts allows us to question more.

Students in grades 3 and 4 will focus on fractions as part-to-whole relationships and as measures. In grade 5, students will then be exposed to fractions as division and as operators. It is not until grade 6 (middle school), that children are exposed to fractions as ratios. It's also important to note that in elementary school, children aren't formally exposed to fractions until third grade. Now, that doesn't mean they have never heard the terms *half*, *third*, *fourth*, and such. Children are exposed to fractions in real-life contexts and are certainly around the language (e.g., "it's a *quarter* past the hour").

However, the formal abstract notation where we write a fraction as $\frac{a}{b}$, where a is a whole number and b cannot be zero should not be taught until students are conceptually ready to internalize that notation. In grades 1 and 2, students will **partition**, or divide, shapes into equal-size parts and identify those parts as *fourths* or *halves*, but there is really no reason for children to write the abstract notation before they've developed a deeper understanding of fractions.

Table 7.1. Common Interpretations of Fractions (Using $\frac{4}{5}$ as an Example)

Fractions Are . . .	*Understanding*	*How to Help*
Part-Whole relationships (Grade 3+)	One whole partitioned, or cut, into equal-size pieces Example: $\frac{4}{5}$ represents 4 pieces of one-fifth size	When children first learn about fractions, they often learn about the part to whole relationships before other interpretations of fractions. Help children connect this understanding to multiplication by seeing that the numerator represents the number of pieces and the denominator represents the size of each piece.
Measures (Grade 3+)	Because a fraction is a number, it can be located on a number line. Examples: $\frac{4}{5}$ inch on a ruler; or $\frac{4}{5}$ of an hour past 2 on a clock shows the minute hand at 48. In these cases, $\frac{4}{5}$ is a measure of 4 one-fifth size units from the starting point (zero).	Call the fraction by its name: *four-fifths*. Avoid language such as "four *over* five" which can make kids think fractions are two separate numbers. Look for opportunities to help your child see that $\frac{4}{5}$, for example, represents a number on a number line.
Operators (Grade 5+)	A fraction can be used to "operate" on a quantity. In other words, a fraction can stretch or shrink the size of a number. Example: A model airplane is $\frac{4}{5}$ the size of an actual airplane.	Without context, fractions can be interpreted in many different ways. Encourage your child to communicate precisely by labeling units or indicating whether they are shrinking (multiplying by a fraction less than 1) or enlarging a figure (multiplying by a fraction greater than 1).
Quotients (Grade 5+)	$4 \div 5 = \frac{4}{5}$ Four dollars being shared equally among five people results in each person getting $\frac{4}{5}$ of a dollar, or \$0.80, or 80% of a dollar.	The result of dividing is a quotient, which represents a number. When reviewing basic division facts with your youngster, discuss how division and fractions are related.

So, this chapter really builds upon the knowledge that students in grades 3 and above rely upon. If you have a child who is not yet in third grade, this doesn't prevent you from reading this chapter! Instead, we hope you read the chapter to prepare yourself for how you will shift your language to encourage understanding, rather than a focus on answer-getting.

Did You Know?

In the past, instruction of fractions in schools focused on the outcomes—memorizing procedures so that students can successfully operate with fractions. However, being a good mathematical thinker is not based on how quickly someone can produce a quantitative answer. It is more important that mathematical thinkers understand processes and have multiple pathways to a solution than to lack the ability to problem solve.

THE WHOLE MATTERS

Remember back in school when you learned that $\frac{3}{4}$ is greater than $\frac{1}{2}$? Is that always true? Can you think of a circumstance in which $\frac{3}{4}$ is actually less than $\frac{1}{2}$? Go ahead. Think about it for a moment. Guess what? It is possible! But . . . how? It's all about the *whole*. When comparing fractions, it is critical that they are compared against the same size whole, otherwise it is like comparing apples to oranges.

If you are comparing $\frac{3}{4}$ of the size of a quarter to $\frac{1}{2}$ the size of a large pizza, then clearly in that instance $\frac{1}{2}$ will be greater than $\frac{3}{4}$. Figure 7.1 shows what we mean. In this case, $\frac{1}{2}$ the area of the pizza is much greater than $\frac{3}{4}$ the area of the coin.

Today's math classrooms spend a lot of time building this deep understanding around the whole to help improve number sense with fractions. Is your child ready for a lesson about the whole? Go to your local convenience store or supermarket and buy a bite-size Kit Kat bar and an XL King Size Kit Kat bar. When you come home, tell your child that they've behaved so well they deserve a *whole* chocolate bar. Your child will be elated! Then, hand them the bite-size bar.

There might be a slight bit of disappointment as they expected a larger bar . . . but just wait until you tell them that as the parent, you don't think you deserve a *whole* and that you will treat yourself to only *half*. They will feel much better about the size they got! But then . . . dramatically take out the King Size Kit Kat bar and break it in half. Their face will be worth the

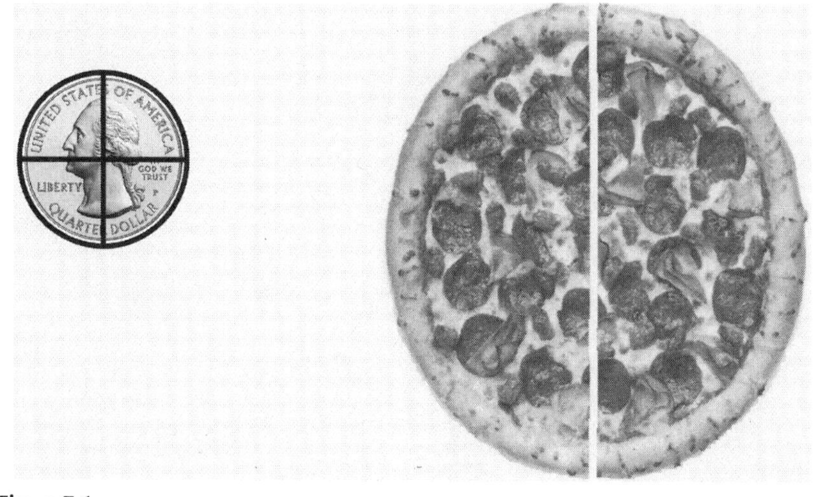

Figure 7.1.

investigation, as they will realize immediately that the *half* you were talking about is much bigger than the *whole* they got! Lead this into a nice discussion about how the unit has to be the same size otherwise comparing fractions is meaningless! Then, we suggest you share some of your half.

MODELS AND REASONING

When discussing fractions throughout this chapter, we will use three main representations, or pictures, to show fractions visually. The models are called a *set model*, an *area or region model*, and a *linear model*. We will also write fractions with symbolic notation, which just means that they will be written as $\frac{a}{b}$ where a is the **numerator** (top number) and b is the **denominator** (bottom number). Take a closer look at these three models in figure 7.2 and notice the real-life examples of these models you might have in your home.

While you don't need to know the models specifically or by name, we wanted to give you a brief overview so you can identify the various models as we use them throughout the chapter.

UNIT FRACTIONS

To really develop a deep understanding of fractions, students must have lots of experience with **unit fractions** from early on. Unit fractions are fractions where the numerator is 1. In a part-whole relationship, the unit fraction

Model	Visual	Examples in Your Home
Area or Region		A clock (quarter of an hour; half an hour) Milk carton (the carton is ¾ full of milk)
Linear	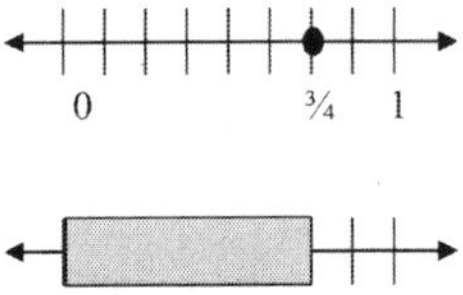 0 ¾ 1 0 ¾ 1	A clock (⅙ of 1 hour is 10 minutes) A ruler or measuring tape (a bed is ¼ inch away from the wall) Measuring cups (you need three ¼ cups to make one serving) Analog scale
Set		Toy sets (a set of automobile toys, some fire trucks and some cars) Dishes (¾ of a set are dinner plates, ¼ are salad plates)

Figure 7.2.

$\frac{1}{5}$	$\frac{1}{5}$	$\frac{1}{5}$	$\frac{1}{5}$	$\frac{1}{5}$	= 1

Figure 7.3.

represents one part of a whole that has been partitioned into equal-size parts. For example, $\frac{1}{5}$ is one part of five equal-size parts. This means that five copies of that one part ($\frac{1}{5}$) is one whole ($\frac{5}{5} = 1$). (See figure 7.3.)

The development of unit fractions is critical, as now students have the language necessary to move forward with understanding **nonunit fractions**.

Did You Know?

Ancient Egyptians only used unit fractions! They would not write $\frac{4}{5}$ as a fraction the way we would but, instead, as a sum of unit fractions! In fact, to make matters more challenging, they would not repeat any unit fractions! For example, they would write either $\frac{1}{2} + \frac{1}{5} + \frac{1}{10}$ or $\frac{1}{2} + \frac{1}{4} + \frac{1}{20}$ to represent $\frac{4}{5}$, but they would never write $\frac{4}{5}$ or $\frac{1}{5} + \frac{1}{5} + \frac{1}{5} + \frac{1}{5}$. The only exception was $\frac{2}{3}$. which was allowed to be written as $\frac{2}{3}$.

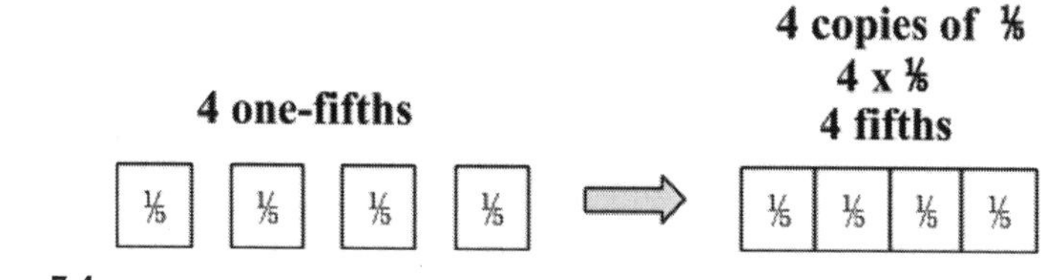

Figure 7.4.

NONUNIT FRACTIONS

Nonunit fractions are fractions that are not written with a numerator of 1. Examples of nonunit fractions are $\frac{2}{3}$, $\frac{5}{7}$, and $\frac{6}{4}$, to name a few. Because students have spent most of grade 3 focusing on unit fractions, they will talk about nonunit fractions in third grade by describing the quantity of unit fractions; for example, they will talk about the fraction $\frac{4}{5}$ as "4 *one-fifths*." As students enter grade 4, they will connect the language of "4 *one-fifths*" to the operation of addition by writing $\frac{4}{5}$ as the sum of unit fractions (*e.g.*, $\frac{1}{5} + \frac{1}{5} + \frac{1}{5} + \frac{1}{5} = \frac{4}{5}$). By the end of fourth grade, students will begin to draw a connection to the operation of multiplication and realize that 4 *one-fifths* is the same as writing $4 \times \frac{1}{5} = \frac{4}{5}$ (see figure 7.4).

COMPARING FRACTIONS

What does it mean to compare fractions? Comparing quantities is all about identifying whether a number is greater than, less than, or equal to another number. When we learned about how to compare fractions as kids, we were often taught to just find a common denominator, or convert the fraction to a decimal, or do some fancy trick that half the time we couldn't remember after leaving the classroom. Perhaps you shared a similar experience. Today, just like with all other aspects of math, kids are taught to *reason* and *think critically* rather than just compute numbers.

When asked to compare fractions, students over the course of grades 3 and 4 will learn a variety of strategies. The most important takeaway for children needs to be that they are in charge of deciding which strategy to use and when. Just knowing a bunch of strategies won't be helpful if the strategies are only helpful in certain situations.

To ensure that kids have a deep understanding of comparing fractions, we need to start helping them make sense of the fractions first. This idea is known as *benchmarking*. A benchmark is a point of reference that can be used to compare things. Alternatively, you might hear your child say "friendly number" or "landmark number." These terms are interchangeable and focus

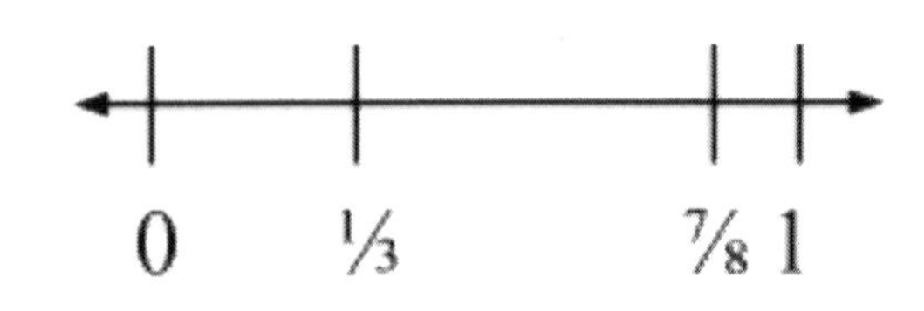

Figure 7.5.

on the idea that in order to compare fractions, we must be actively paying attention to where that fraction is located on a number line.

For example, let's take a look at two fractions: $\frac{1}{3}$ to $\frac{7}{8}$. Now, as adults, comparing these two fractions seems silly—obviously $\frac{7}{8}$ is greater than $\frac{1}{3}$. However, kids who are just learning how to compare fractions might not find this as easy as we do! But, they might be able to make sense of this by comparing both fractions to **benchmarks** on the number line. A child might say that $\frac{1}{3}$ is closer to 0 or $\frac{1}{2}$, whereas $\frac{7}{8}$ is closer to 1, which then tells them that $\frac{7}{8}$ is greater (see figure 7.5).

Similarly, if comparing $\frac{5}{9}$ to $\frac{3}{7}$, a student might think about how close each fraction is to $\frac{1}{2}$. A child might first think, "What is half of 9?" and "What is half of 7?" to determine whether the numerator of the fractions are more or less than the halfway point. Since half of 9 is 4.5 (said as *four and one-half, four and five tenths*, or as *four point five*), we know that $\frac{5}{9}$ must be greater than $\frac{1}{2}$ because $\frac{5}{9} > \frac{4.5}{9}$. Since half of 7 is 3.5, we can see that $\frac{3}{7}$ must be less than $\frac{1}{2}$ because $\frac{3}{7} < \frac{3.5}{7}$. You will see the strategy of benchmarking come up again later in this chapter as we show how to operate (add, subtract, multiply, and divide) with fractions.

Once students have a strong sense of fractions and their placement on the number line, there are four main strategies that students will learn for comparing fractions using reasoning skills. We list the four strategies below and provide examples of each.

Common Denominators

Thinking back to earlier in the chapter, denominators are the bottom numbers when a fraction is written in *fraction notation* (e.g., $\frac{a}{b}$). In math, a common denominator means that the bottom numbers of fractions are the same. For example, if we compared $\frac{3}{5}$ to $\frac{2}{5}$, we'd be comparing fractions that have common denominators. As adults, this type of comparison seems like a no brainer, but remember, kids don't have all the lived experiences and knowledge we as adults do, so this can be a challenging concept.

To compare fractions when they have a common denominator, students learn to justify by typically using a part-whole relationship context. Figure 7.6 gives a visual of an area model showing $\frac{3}{5}$ and $\frac{2}{5}$.

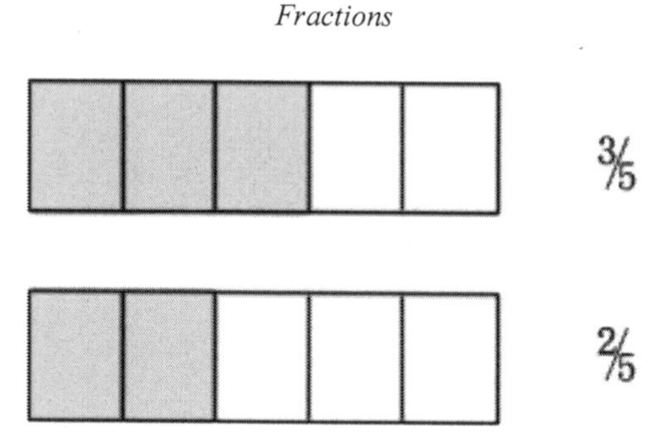

Figure 7.6.

The denominator tells us the total number of equal-sized parts, in this case 5. The numerator tells us how many parts we have. So, for the first whole, we have 3 parts, and for the second whole we have 2 parts. Since 3 > 2 and we are comparing the same size whole, then $\frac{3}{5} > \frac{2}{5}$.

Common Numerators

Similar to common denominators, common numerators means the top number of a fraction written in fraction notation (e.g., $\frac{a}{b}$) is the same as the numerator of another fraction. For example, if we compared $\frac{3}{5}$ to $\frac{3}{8}$, we'd be comparing fractions that have common numerators. Again, thinking of a part-whole relationship, the numerators represent the number of parts, whereas the denominators represent the size of each part. Figure 7.7 gives a visual of $\frac{3}{5}$ and $\frac{3}{8}$.

Without using a visual, kids who are able to reason mathematically will know and be able to explain that $\frac{3}{5} > \frac{3}{8}$ because if they have the same number of pieces, then they can just look at the denominator and compare the size of the pieces. This ties back nicely to the deep understanding of fractions. Essentially, kids might say to themselves, "Which is greater: 3 *one-fifths* or 3 *one-eighths*?"

It makes sense when kids form a mental image: The more we partition, or cut, a whole, the smaller the size of the parts. So, in a realistic context, if we cut a pizza into five slices, we will have bigger slices than if we cut the same pizza into more slices, such as eight pieces. Since they are comparing the same number of slices, it just comes down to the size of each slice.

Residual Thinking

Residual thinking is a fancy term for comparing what remains, rather than what we are given. This is a great strategy that kids can use when the fractions are unit fractions away from a whole, one-half, or another

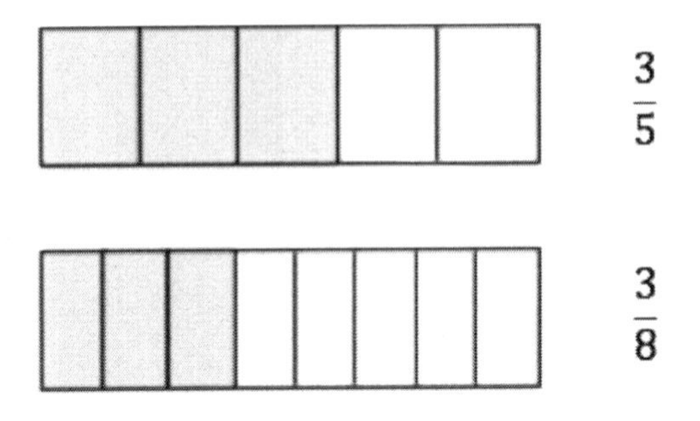

Figure 7.7.

easy-to-work-with number. To showcase this strategy, let's imagine we are comparing $\frac{7}{8}$ to $\frac{6}{7}$. These fractions are so close to each other that if we drew a visual with pencil and paper, we might be imprecise and make an error in our comparison.

A different way to think about comparing these two fractions is to think about the missing parts. For example, $\frac{7}{8}$ is $\frac{1}{8}$ away from one-whole. Similarly, $\frac{6}{7}$ is $\frac{1}{7}$ away from one-whole. Using this understanding, we now can just compare the unit fractions $\frac{1}{8}$ to $\frac{1}{7}$. Thinking back to our common numerator strategy, because we have the same numerators, we can just look at the denominator to focus on the size of the pieces. In this case, the greater the denominator, the smaller the piece. Knowing that, we can determine that $\frac{7}{8}$ is greater than $\frac{6}{7}$ because it is only $\frac{1}{8}$ away from one-whole, whereas $\frac{6}{7}$ is $\frac{1}{7}$ away from one-whole (see figure 7.8).

As you can see from figure 7.8, because eighths are smaller than sevenths, $\frac{7}{8}$ is closer to 1 than $\frac{6}{7}$ is.

Equivalent Fractions

As students become savvier with fractions, they will begin to look for ways in which fractions that are being compared can be made to look similar. This is when creating an **equivalent fraction** might come in handy. Equivalent fractions are fractions that share the same value but have different numbers of pieces to the same whole. For example, $\frac{3}{6}$ is equivalent to $\frac{1}{2}$ because both fractions represent the same amount of area, but the whole is partitioned into more pieces when considering the sixths. (See figure 7.9.)

Using their understanding of equivalent fractions, students can compare, for example, $\frac{9}{11}$ to $\frac{3}{5}$ by looking for relationships between the numerators or the denominators. In this case, noticing a relationship between the numerators could prove helpful because a child might think of $\frac{3}{5}$ as $\frac{9}{15}$ (equivalent

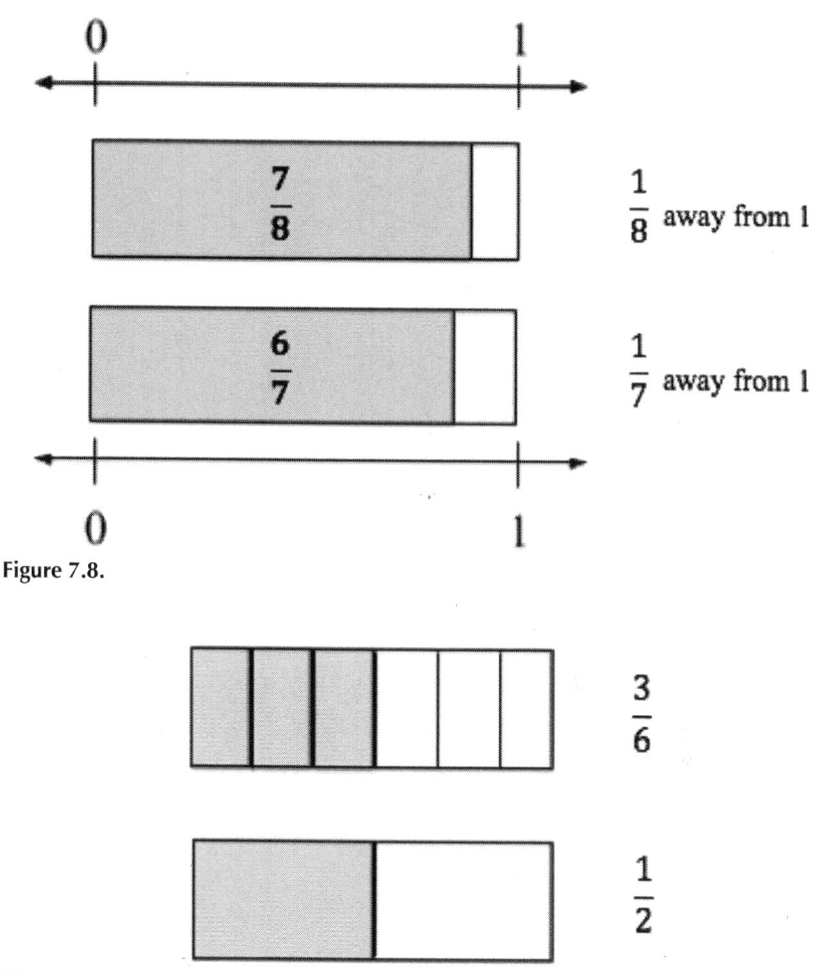

Figure 7.8.

Figure 7.9.

fractions). This helps them then compare *common numerators* (e.g., $\frac{9}{11}$ compared to $\frac{9}{15}$). Using the common numerator strategy, we know to just compare the denominators, which tell us the size of the pieces. Since $\frac{9}{15}$ is cut into more pieces, the size of the pieces must be smaller than $\frac{9}{11}$. Therefore, $\frac{9}{11}$ is greater than $\frac{3}{5}$.

Of course, there are more ways to compare fractions; as students become more familiar with decimals in fifth grade and fractions as division, they might convert fractions to decimals to compare them. Ultimately, for the purpose of this book, we wanted to focus on the main strategies students will be using that focus on mental reasoning and thinking critically.

OPERATIONS WITH FRACTIONS

When we think about operating with fractions, we want you to draw connections to the work we did in chapter 5 (addition and subtraction of whole numbers) and chapter 6 (multiplication and division of whole numbers) since whole numbers and fractions are both numbers. To help you visualize this, we will use whole number operations and models throughout this chapter.

Adding and Subtracting Fractions

Just like with whole numbers, when we think of adding and subtracting numbers, we might visualize different contexts that help us make sense of the operations. In the beginning, to understand adding fractions, it's important to remember that whole numbers are made up of sums of wholes (e.g., $2 = 1 + 1$), and similarly, fractions are made up of sums of unit fractions (e.g., $\frac{2}{3} = \frac{1}{3} + \frac{1}{3}$). This idea is foundational for students to be able to compose and decompose fractions to operate mentally.

Be sure when working with your child at home you don't forget about fractions *greater than one*. Growing up, you may have called these *improper fractions*, and while they are still called that, we suggest just saying *fractions greater than one*, because really, that's what they are. This, then, allows students to make the initial connection that if the numerator is greater than the denominator, then the value is greater than one. For example, $\frac{4}{3}$ is a fraction greater than one and can be written as as sum of unit fractions (e.g., $\frac{4}{3} = \frac{1}{3} + \frac{1}{3} + \frac{1}{3} + \frac{1}{3}$).

Even just seeing the sum of unit fractions, kids can begin to make connections to $1\frac{1}{3}$ because they see $\frac{1}{3} + \frac{1}{3} + \frac{1}{3}$ as 1 and then there's an extra $\frac{1}{3}$, which makes it $1\frac{1}{3}$. Rewriting fractions greater than one as *mixed numbers* (or whole numbers with fractions, such as $1\frac{1}{3}$) now has meaning and is not just a trick for kids to memorize. Thinking about the language, $\frac{4}{3}$ is really just four groups of *one-third*. In fact, we can combine three *thirds* to make 1 and have $\frac{4}{3} = \frac{3}{3} + \frac{1}{3}$. Look at figure 7.10 to see why it makes sense!

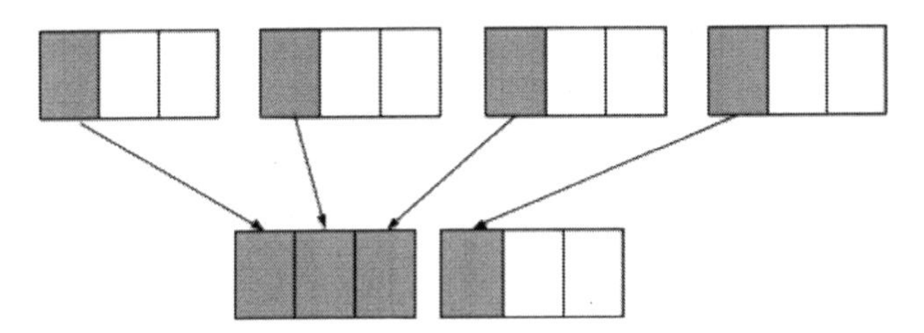

Figure 7.10.

All of this becomes very useful when we begin adding fractions with *common denominators*. Thinking back to comparing fractions, fractions with common denominators have the same denominator. So, take, for instance, $\frac{1}{5} + \frac{3}{5}$. Using the same logic as before, encourage your child to say, "*one-fifth* plus *three-fifths*" and help them notice that *fifths* is acting as a unit here. Since we are adding the same unit, we can just add the numerators to get $\frac{4}{5}$. Notice how *one-fifth* plus *three-fifths* is *four-fifths*. Just like if we had *one apple* and added *three apples* we would have *four apples*, common denominators allow us to add fractions that are in the same form. Unfortunately, if taught procedurally, this will be the beginning of where kids tend to start misusing rules with fractions.

Lots of kids who aren't thinking mathematically will accidentally add the numerators *and* add the denominators (e.g., $\frac{1}{5} + \frac{3}{5} = \frac{4}{10}$) and get the wrong answer. If your child does that, this is a critical place for you to ask a question back, rather than tell them what they did wrong and show them the right way. Remember, the goal is learning to be able to apply it in the future on their own! What question might you ask your child if they added the numerators *and* the denominators? Some questions we'd suggest are:

- Does your answer make sense?
- Can you draw a picture to show me what $\frac{1}{5}$ looks like when added to $\frac{3}{5}$?
- If the question were asking $\frac{1}{2} + \frac{1}{2}$ would you get $\frac{2}{4}$? Would that make sense?
- Is $\frac{3}{5}$ more or less than $\frac{1}{2}$? Follow-up question: If it's more, then when you add $\frac{1}{5}$ will your answer be more or less than $\frac{1}{2}$? Is $\frac{4}{10}$ more or less than $\frac{1}{2}$?

Again, the point is to always question your child's thinking, even if their answers are correct, to ensure they can make sense of the problem and that they are using logical thinking and not just step-by-step procedures.

Just like adding fractions with common denominators, subtracting fractions with common denominators, or "like denominators," focuses on the language. If we were to find $\frac{4}{5} - \frac{2}{5}$, we could think of the denominators like units. *Four-fifths* minus *two-fifths* is *two-fifths*, just like 4 apples minus 2 apples is 2 apples. We encourage parents to help kids make sense of fractions at home by focusing on the language of the math and supporting this understanding with visuals.

As students move on to adding and subtracting fractions with denominators that are different, or unlike denominators, we want to help them use what they have already learned and apply it. Namely, we want them to use critical thinking rather than procedures.

Say, for instance, a child were asked to add $\frac{3}{4} + \frac{5}{8}$. Using our strategy of decomposition of *whole numbers* (chapter 5, addition and subtraction of whole numbers), we can break apart $\frac{5}{8}$ into $\frac{2}{8} + \frac{3}{8}$. Why might we want to think about $\frac{5}{8}$ in that manner? Well, because we know that $\frac{2}{8} = \frac{1}{4}$ and $\frac{1}{4} + \frac{3}{4} = 1$. This makes our lives so much easier because the original problem can be thought of as $\frac{3}{4} + \frac{2}{8} + \frac{3}{8}$. If we rewrite this number sentence as $\frac{3}{4} + \frac{1}{4} + \frac{3}{8}$, you can quite clearly see that the sum is $1\frac{3}{8}$. If you look at figure 7.11, you can see this same process unfold on a number line where we decompose the $\frac{5}{8}$ to help us "get to the whole."

Once your child has some fluency with decomposing fractions to aid in the addition and subtraction process, it's helpful to then encourage them to draw area models to help them visualize common denominators. Let's take the same problem from before and look at it through the use of area models. If we want to add $\frac{3}{4} + \frac{5}{8}$ using area models, we have to first start out by drawing two area models: one representing $\frac{3}{4}$, and one representing $\frac{5}{8}$ where the wholes are the same (see figure 7.12).

At this point, it should be visually obvious that these two wholes are partitioned into different sized parts. We know to add fractions, we must have

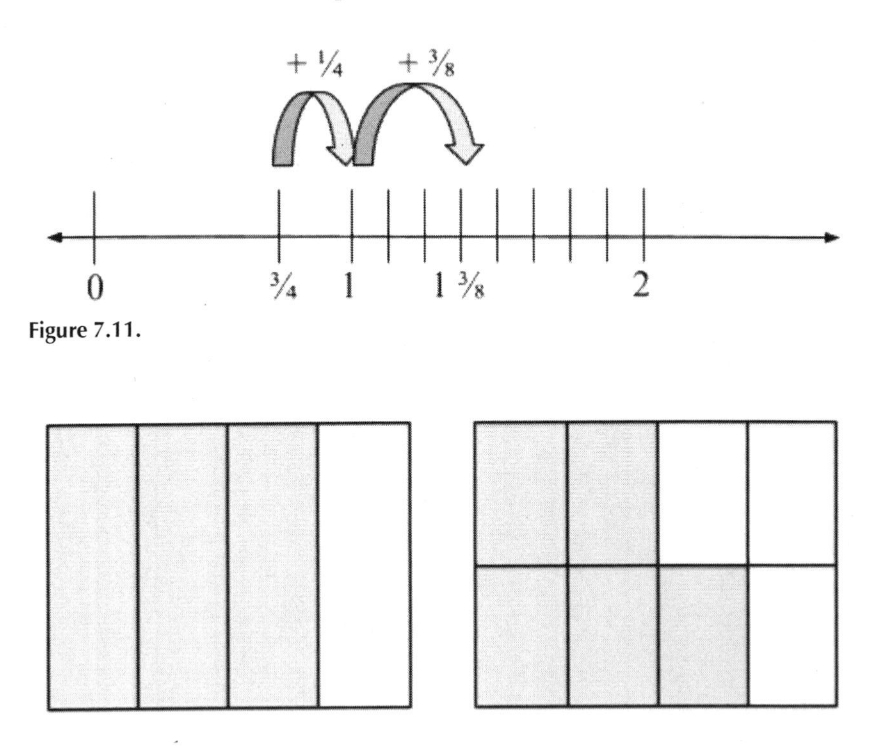

Figure 7.11.

Figure 7.12.

the same sized parts. There are many ways we can create same sized pieces, but we will create eighths by partitioning the fourths in half with a horizontal line (see figure 7.13).

Now, we have the same sized wholes and the same sized parts, so adding the parts together is now much easier. We can fill the two empty boxes of the first whole using two of the parts from the second whole. Then, our answer will be really visible (see figure 7.14).

It is incredibly clear from figure 7.14 that we have $1\,\frac{3}{8}$, and it's also easy to see that we could alternatively say we have $\frac{11}{8}$ because we literally have *eleven-eighths*.

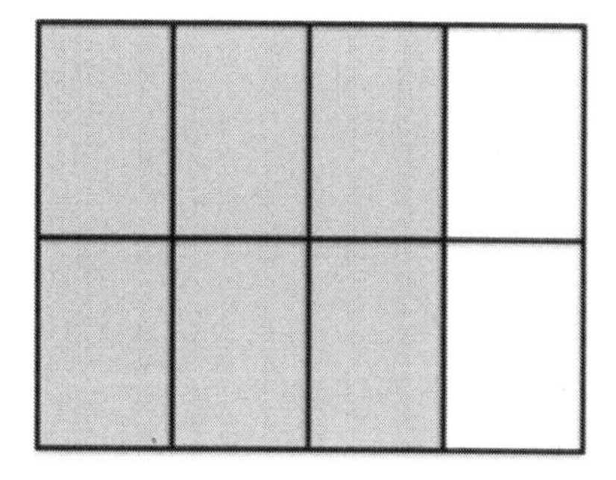

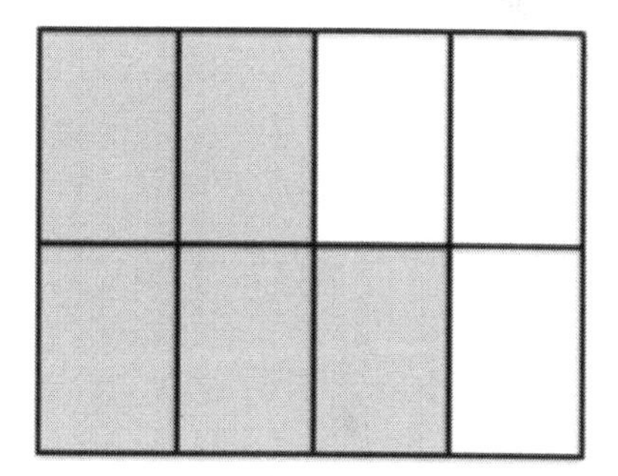

$$3/4 + 5/8$$

Figure 7.13.

$$3/4 + 5/8 = 1\ 3/8$$

Figure 7.14.

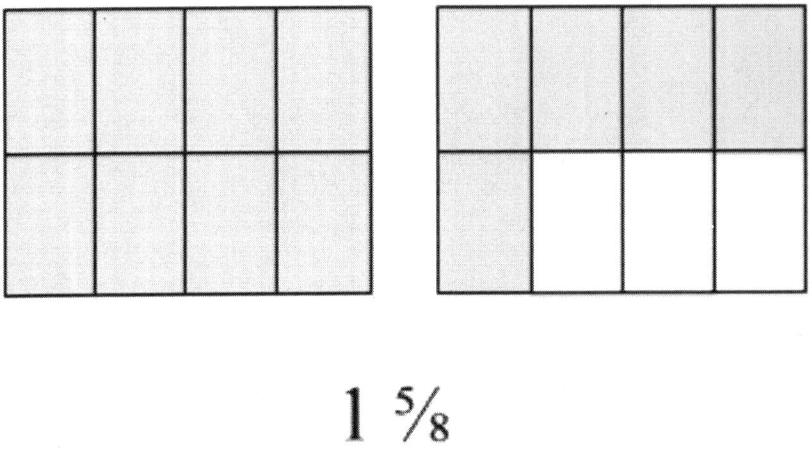

$1\,{}^{5}\!/\!_{8}$

Figure 7.15.

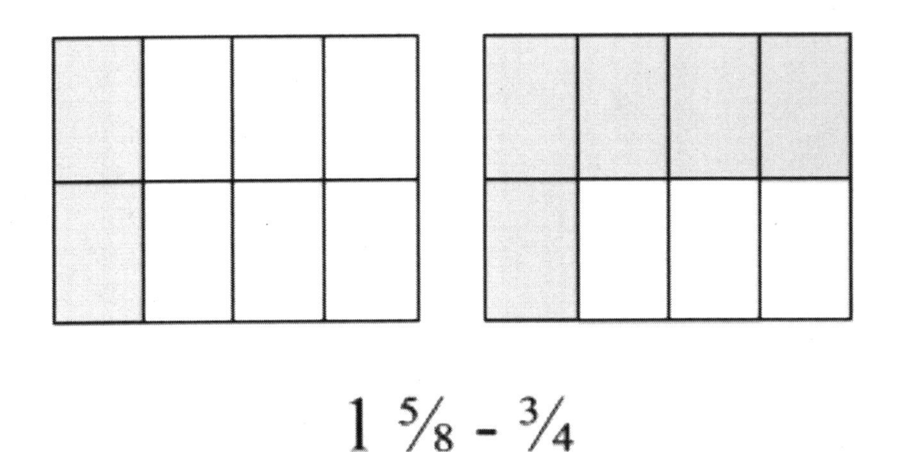

Figure 7.16.

Now let's explore subtraction with unlike denominators and walk through the same strategies and models. Say, for instance, a child were asked to subtract $1\frac{5}{8} - \frac{3}{4}$. Using the strategy of decomposition, we can think of the problem as $1 - \frac{3}{4} + \frac{5}{8}$, We already know how to subtract from 1, so rather than "getting to the whole" like in addition, we can think about "taking from the whole" for subtraction. If we do that, then we have $\frac{1}{4} + \frac{5}{8}$ remaining. Now, we are back to an addition problem, and we are already well equipped to solve a problem like this! Let's look at this using area models to see how the model enhances our understanding.

First, we have to visualize what we have, which is $1\frac{5}{8}$ (see figure 7.15).

Now, we need to take away $\frac{3}{4}$. Looking at the area models, you can see we do not have enough boxes on the second area model to take away $\frac{3}{4}$, which is why we will take it away from the whole, or the area model on the left. By taking away $\frac{3}{4}$, we are left with the following (see figure 7.16).

To make sense of it, let's rearrange the boxes into the first whole so we can see clearly what we have remaining (see figure 7.17).

Using the area model here helped us realize that $\frac{3}{4} = \frac{6}{8}$, though we didn't need to state it. Alternatively, we could have created like denominators from the start and looked at the whole as $\frac{8}{8}$. Figure 7.18 shows how that would have played out.

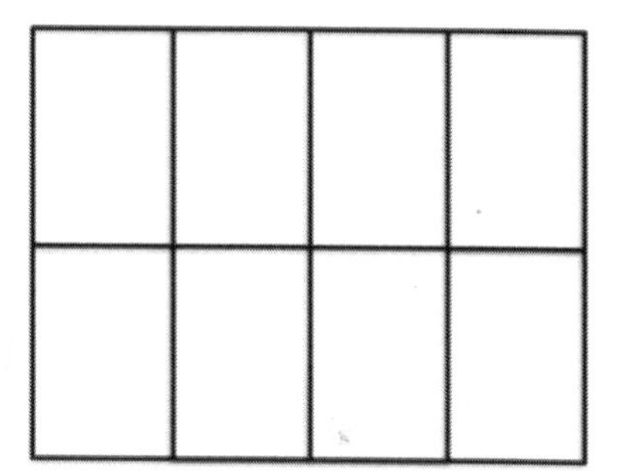

$$1\,\frac{5}{8} - \frac{3}{4} = \frac{7}{8}$$

Figure 7.17.

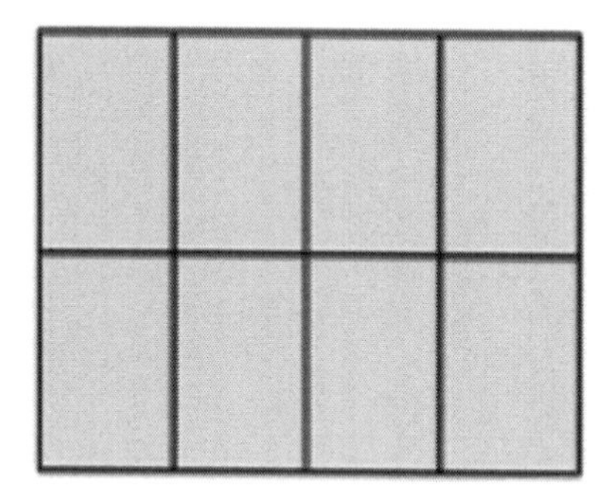

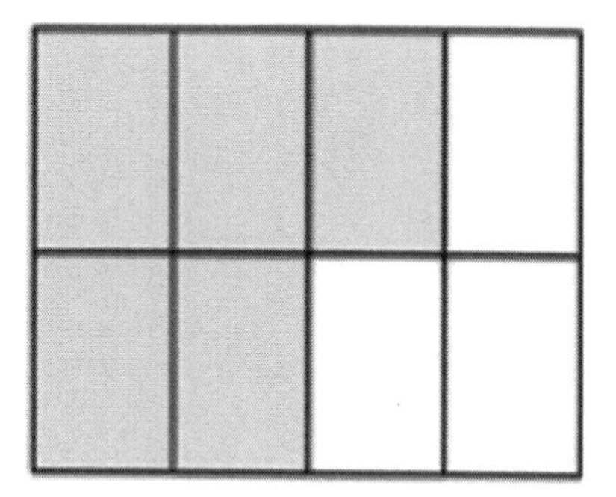

$$\frac{8}{8} + \frac{5}{8} = 1\frac{5}{8}$$

Figure 7.18.

Knowing that *one-whole* in this case is $\frac{8}{8}$, we can say we really have $\frac{13}{8}$ "*thirteen-eighths*." We need to subtract $\frac{3}{4}$, which we can rewrite as $\frac{6}{8}$. Now, $\frac{13}{8} - \frac{6}{8} = \frac{7}{8}$ (see figure 7.19).

While your child will learn to manipulate numbers to "get to the whole" or "take from the whole" and use area models (among other representations) to conceptually begin to make sense of common denominators, they will also eventually learn the procedures for finding a common denominator and then add those fractions together. This strategy will feel most familiar to parents, as it was how most of us learned! Let's look at $\frac{3}{4} + \frac{5}{8}$ through the use of procedures for an example.

To add the fractions with common denominators, we must first create an equivalent fraction to $\frac{3}{4}$ that has a denominator of 8. To do this, we can multiply $\frac{3}{4}$ by $\frac{2}{2}$, because *two-halves* is *1 whole*, and when we multiply $\frac{3}{4}$ by 1, the product is still $\frac{3}{4}$. Even though we've created an equivalent fraction of $\frac{6}{8}$, the value is still the same as $\frac{3}{4}$. Now, we can rewrite $\frac{3}{4} + \frac{5}{8}$ as $\frac{6}{8} + \frac{5}{8}$, which we can solve in the same way we added common denominators before. The result here would be $\frac{11}{8}$ or $1\frac{3}{8}$. Hopefully you can see that visuals are a great way to support understanding.

Multiplying and Dividing

As we showed earlier, students in grade 4 begin drawing connections to multiplying fractions when they realize that instead of repeatedly adding unit fractions together (e.g., $\frac{1}{3} + \frac{1}{3} + \frac{1}{3} + \frac{1}{3} + \frac{1}{3} = \frac{5}{3}$), they can multiply (e.g., $\frac{1}{3} \times 5 = \frac{5}{3}$). This knowledge becomes very helpful as students learn to apply it to more complicated problems. For example, look at the two problems below and

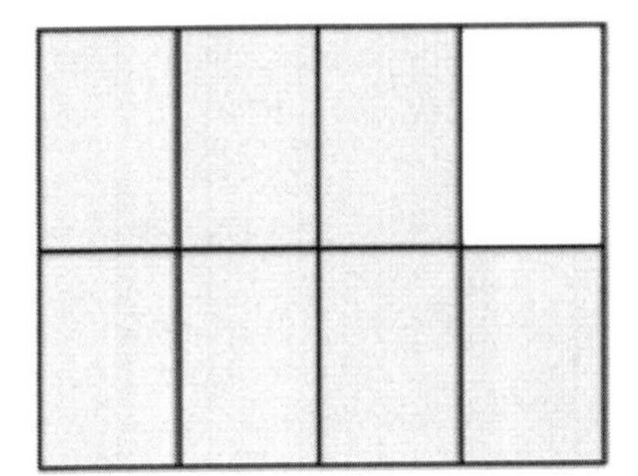

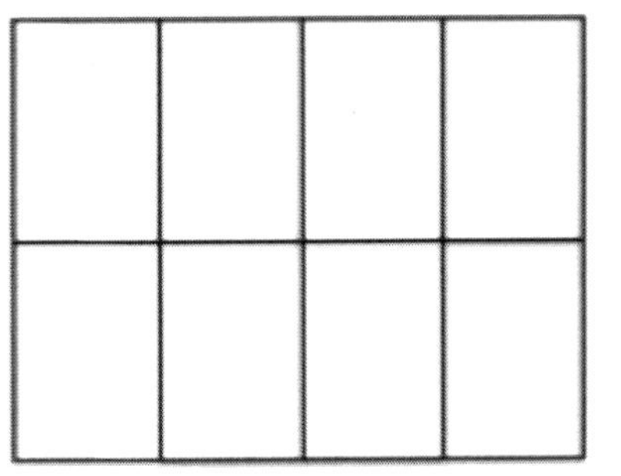

$$1\frac{5}{8} - \frac{6}{8} = \frac{7}{8}$$

Figure 7.19.

notice how the knowledge of the first problem helps students make sense of the second one.

$10 \times \frac{1}{3} =$
$10 \times \frac{2}{3} =$

With $10 \times \frac{1}{3}$, students are familiar with now interpreting this as *10 groups of one-third*, or *ten-thirds*. How does that thinking help them solve the next problem, $10 \times \frac{2}{3}$? Well, this is *10 groups of two-thirds*, which is *two times as many groups* as $10 \times \frac{1}{3}$, which means it must be $\frac{20}{3}$, or *twenty-thirds*. Students can always relate multiplication of a fraction by a whole number to what they would do for a unit fraction by a whole number and then go from there.

As students start to explore fraction by fraction multiplication, they will typically start with unit fractions. For instance, $\frac{1}{2} \times \frac{1}{3}$. Let's read this just as we were reading the whole number by a unit fraction. Instead of $10 \times \frac{1}{3}$, *ten-thirds*, now we are looking at $\frac{1}{2} \times \frac{1}{3}$, or *half-a-third*. Read in this manner, a child might be able to identify that half of one-third is one-sixth ($\frac{1}{6}$). Using an area model, like we did with addition and subtraction, is a great way to visualize this. First, let's look at what $\frac{1}{3}$ looks like (see figure 7.20).

Now, let's partition the thirds in half (see figure 7.21).

We need to identify *one-half* of *one-third*, so our eyes need to focus on just the *one-third* we have (the shaded part) and half of that third is our answer (see figure 7.22).

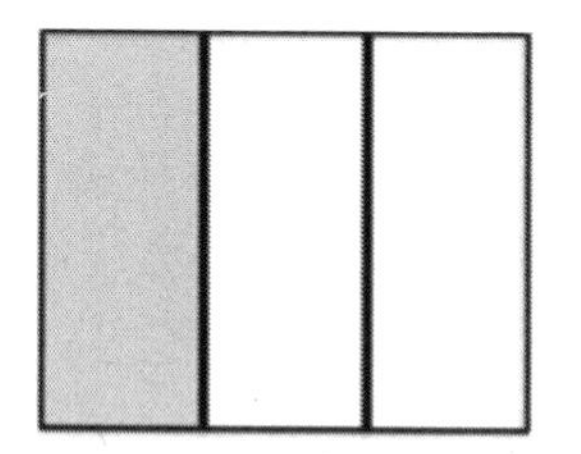

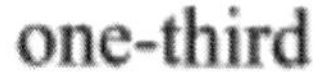

Figure 7.20.

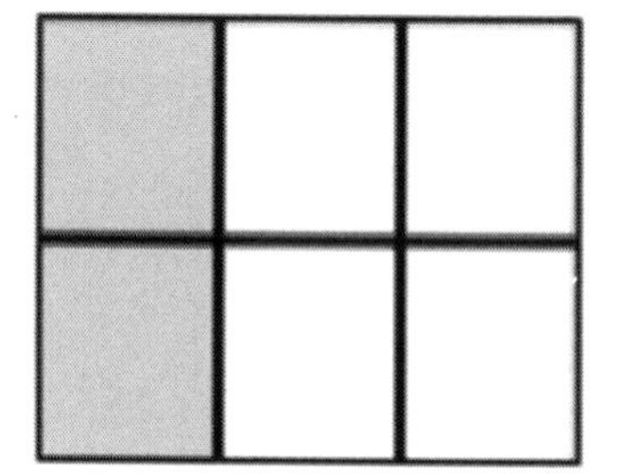

Figure 7.21.

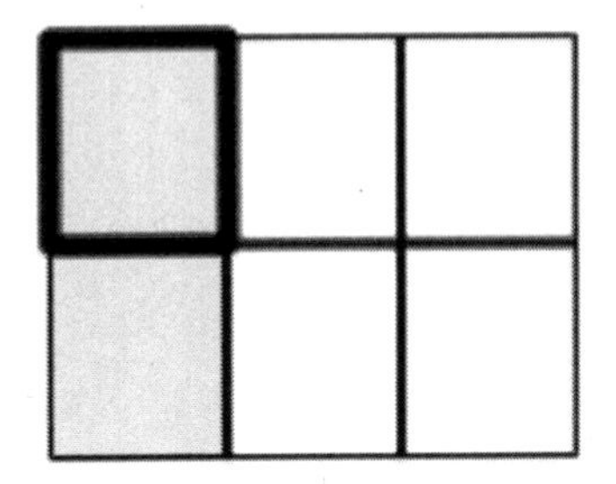

Half of ⅓ is ⅙

Figure 7.22.

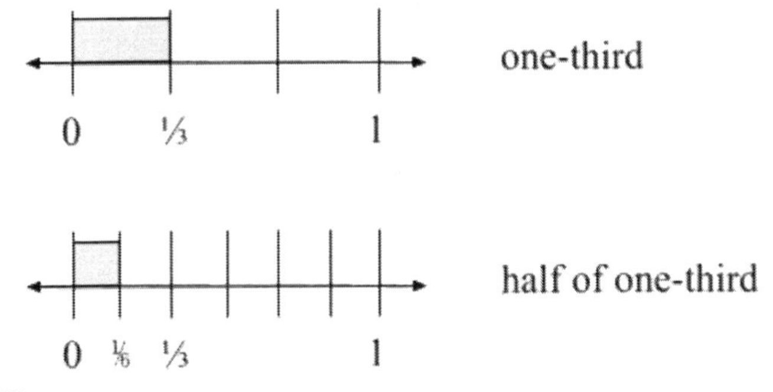

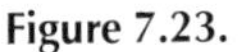
Figure 7.23.

As you can see, the thirds, when cut in half, are now sixths, and we had *two-sixths* shaded (*one-third*) and half of that is *one-sixth*.

Let's explore this same problem using a number line. Remember, we are trying to identify $\frac{1}{2} \times \frac{1}{3}$ (see figure 7.23).

Can you see with the number line that $\frac{1}{3}$ is equivalent to $\frac{2}{6}$? Number lines and area models help us visualize equivalent fractions.

Now let's apply all that you've learned to multiplication of fractions by fractions. Let's solve for $\frac{2}{3} \times \frac{3}{4}$. This time, saying *two-thirds* of *three-fourths* doesn't necessarily help us see the answer any clearer. To build on what we've learned, we can solve for $\frac{1}{3}$ of $\frac{3}{4}$ and then double it because we are actually solving for $\frac{2}{3}$ of $\frac{3}{4}$. Let's explore that with an area model by starting with $\frac{3}{4}$ (see figure 7.24).

To find $\frac{1}{3}$ of $\frac{3}{4}$, we need to partition the fourths also into thirds (see figure 7.25).

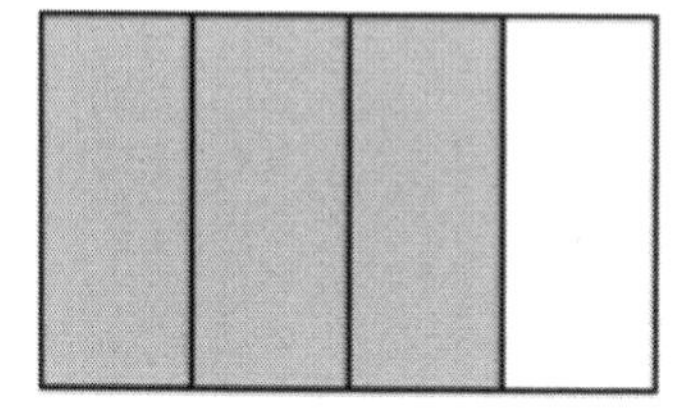

Figure 7.24.

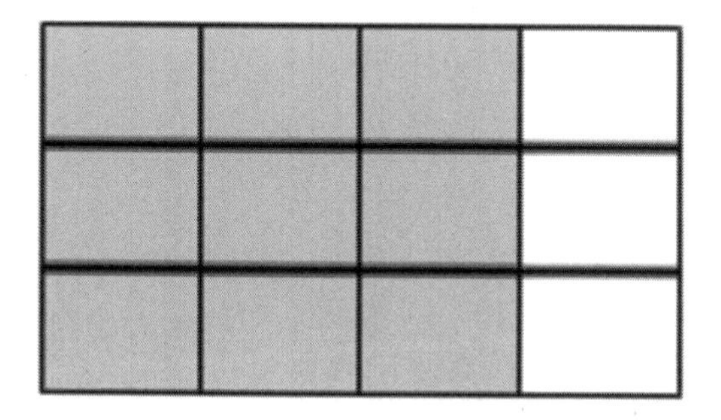

Figure 7.25.

Now, you can see we have *twelfths*. We need to find $\frac{2}{3}$ but can start with $\frac{1}{3}$ and double it (see figure 7.26).

From this, we can see that $\frac{1}{3}$ of $\frac{3}{4}$ is $\frac{3}{12}$. Doubled, $\frac{2}{3}$ of $\frac{3}{4}$ would be $\frac{6}{12}$, or *one-half* (see figure 7.27).

It's important for students to notice that the **product** is not greater than the factors. Because we are finding a part of a part, we are actually looking for a smaller piece, not a bigger piece.

Let's look at $\frac{2}{3} \times \frac{3}{4}$ on a number line for comparison to the area model (see figure 7.28).

The number line shows beautifully how $\frac{2}{3} \times \frac{3}{4} = \frac{1}{3}$.

It's not until grade 5 that students are exposed to fractions as division (e.g., $5 \div 3 = \frac{5}{3}$) and, to be specific, a very limited version. Students in grade 5 should only be dividing a non-zero whole number by a unit fraction or a unit fraction by a non-zero whole number (e.g., $4 \div \frac{1}{2}$ or $\frac{1}{3} \div 5$). Division of non-unit fractions by nonunit fractions does not occur until grade 6 (e.g., $\frac{2}{3} \div \frac{3}{4}$).

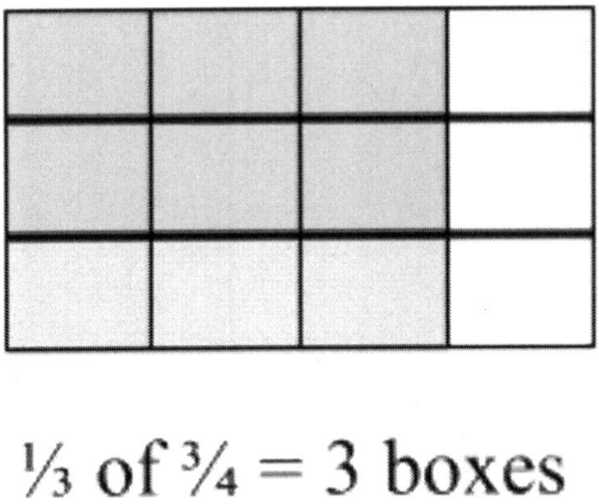

$\frac{1}{3}$ of $\frac{3}{4}$ = 3 boxes

Figure 7.26.

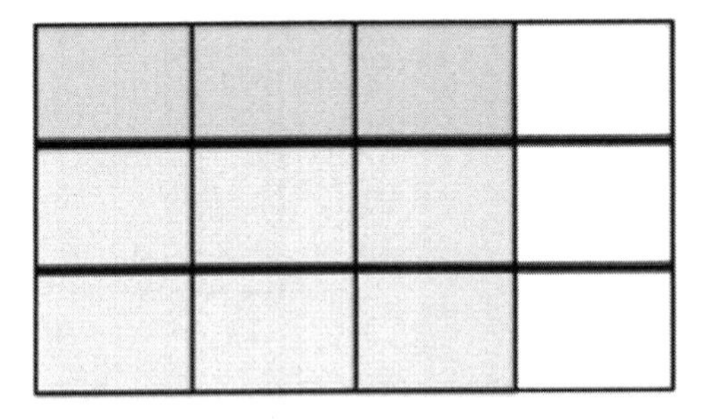

$\frac{2}{3}$ of $\frac{3}{4}$ = 6 boxes

Figure 7.27.

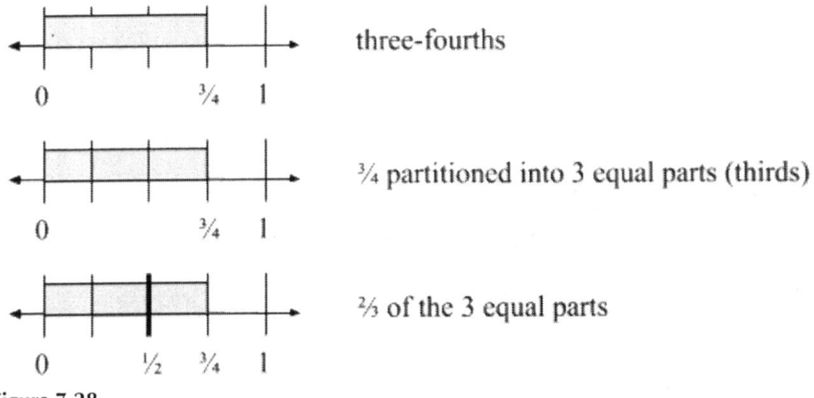

Figure 7.28.

What's beautiful about the progression of the way math is taught is that the concepts connect so nicely together. Let's take $\frac{1}{5} \div 3$ as an example.

In grade 4, students saw that $\frac{5}{3}$ was the same as $5 \times \frac{1}{3}$. Extending this thinking, now we think of $\frac{5}{3}$ as division and recognize that $5 \div 3 = 5 \times \frac{1}{3}$. What do you notice about this equation? Kids in grade 5 explore what it means for $5 \div 3$ to be the same as $5 \times \frac{1}{3}$. With enough repeated practice and exposure, children will recognize on their own that dividing by a number (in this case, 3) is the same as multiplying by the *reciprocal*, a number when multiplied gives 1 (in this case, $\frac{1}{3}$). Making this connection leads perfectly into understanding how to find $\frac{1}{5} \div 3$.

Once students have generalized this rule, now solving $\frac{1}{5} \div 3$ is easy because they realize that $\frac{1}{5} \div 3 = \frac{1}{5} \times \frac{1}{3}$. If this connection is not made intrinsically, then students will most likely just be taught procedures and ultimately mix up rules and misapply them.

Another way students might visualize division of unit fractions is using an area model. Let's look at two problems using the model to help us see why we get the answers we do. To start, let's examine $5 \div \frac{1}{3}$. Thinking back to chapter 6 (multiplication and division of whole numbers), we can think of the various ways to interpret a division problem. Creating story context becomes very helpful here.

Imagine the story is this: There are 5 cups of cereal in a box. The serving size is $\frac{1}{3}$ cup. How many $\frac{1}{3}$ cup servings are there in 5 cups of cereal?

Notice how we started with 5 wholes and were looking for how many groups of $\frac{1}{3}$ could fit into 5 (see figure 7.29).

Now, let's look at the other division problem, $\frac{1}{3} \div 5$, to see how it differs. In the last story context, we used the division as repeated subtraction, so in this story problem, we will use a division as sharing context. Imagine there is $\frac{1}{3}$ of a pound of chocolate and there are 5 people. How much will each person get if they share the chocolate equally?

Each person will get $\frac{1}{15}$ of a pound of chocolate (see figure 7.30).

While fractions can appear intimidating, they are really just numbers, a way to extend our number system beyond 1, 2, 3, and more. Fractions are everywhere, and the more you make your child aware of that, the more important they'll believe they are.

Do you watch sports? Think about all the fractions that occur during any game! There are four quarters in a basketball game, so after 45 minutes, how many fourths have been played? What are the fractions that represent the half-times of each type of sport? Are they all the same? Say the batting averages (written as decimals) aloud and you've got fractions!

Beyond sports, fractions are really everywhere! As your child loses their teeth, have them keep track of the fraction of teeth remaining or lost! Have

5 cups of cereal

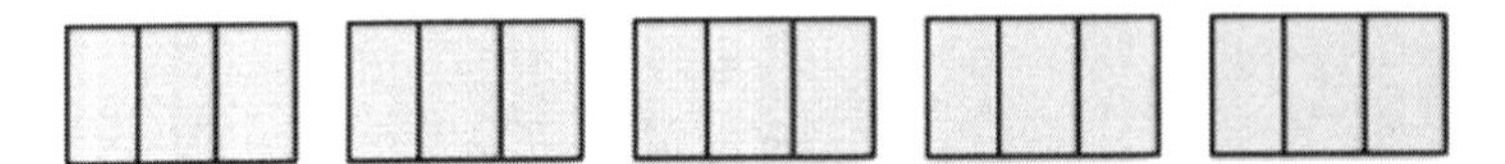

5 cups of cereal
partitioned into thirds

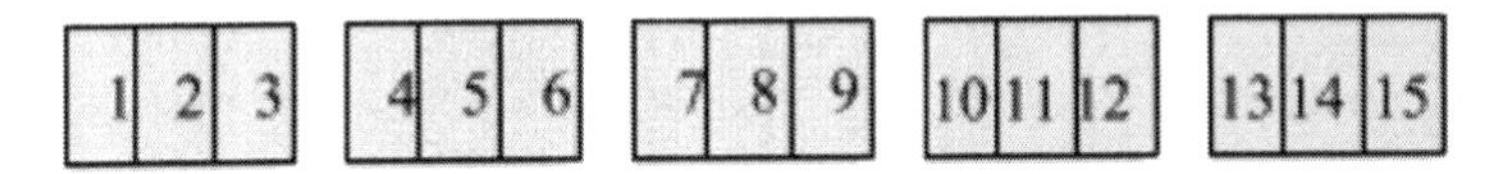

There are 15 servings of ⅓ cup

Figure 7.29.

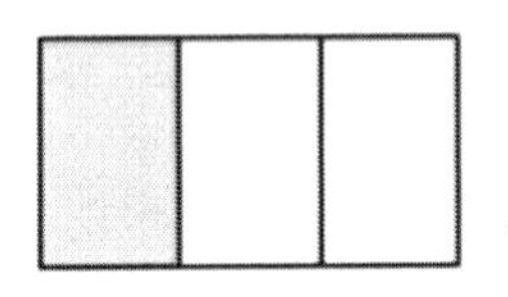

⅓ lb of chocolate

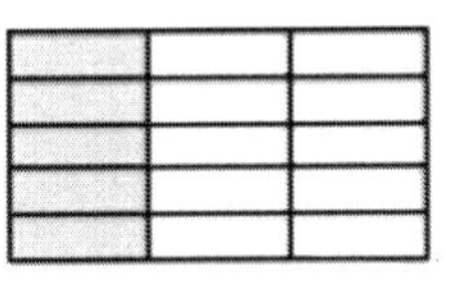

⅓ lb of chocolate
split evenly between
5 people

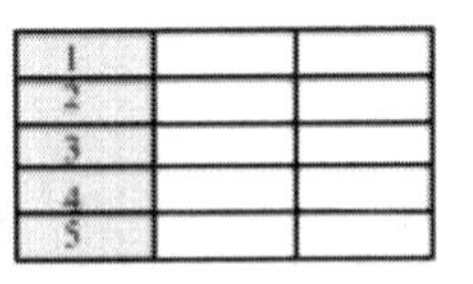

Each person gets
1/15 of a pound

Figure 7.30.

your child look for fractions on the highway ($\frac{1}{2}$ a mile until the exit; $\frac{1}{4}$ mile until the exit). Let your kids know when you're at $\frac{1}{2}$ or $\frac{1}{4}$ tank of gas and show them how you know. When doing handiwork around your home, show your child that wrenches, for example, are named by fractions ($\frac{1}{4}$-inch wrench; $\frac{1}{2}$-inch wrench). Overall, the more you pay attention to the fractions and math around you, the more your child will begin to do the same.

Chapter 8

Preparing Students for the Future

Beyond Elementary Mathematics

> "Good mathematics is not about how many answers you know. . . . It's how you behave when you don't know."
>
> —Author unknown

Mathematics is more critical today than ever. There are several factors that are contributing to this shift, including higher mathematical rigor on state-mandated tests, scholarship money being tied to standardized test scores, and the globalization of the workforce requiring strong mathematical and logic skills. So, it is critical that parents become actively involved in helping their child(ren) become mathematically proficient. Your attitude about learning math will directly impact your child's attitude and success when it comes to mathematics, so be sure to speak positively about math.

Middle school years are very important for a child's algebraic development, which is the foundation of high school mathematics. At the same time, children are often moving from enjoying math—and school for that matter—to disliking the class. There are a variety of factors in play here, including the mindset of the kids, but the lack of conceptual understanding really impacts many of the middle school issues. The majority of the concepts covered in middle school are built from their understanding of topics covered during their elementary years.

High school standards require students to begin to apply mathematical principles in terms of real-world issues and challenge students to begin justifying answers and using higher-level critical thinking skills. The eventual goal is for all students to be college and/or career ready upon leaving high school.

The skills they begin learning in elementary schools are built upon all the way through high school by virtue of a *vertical alignment*. In the most basic sense, vertical alignment is where what students learn in one lesson, course, or grade level prepares them for the next lesson, course, or grade level. This alignment fills the learning gaps from one year to the next. It allows there to be a consistency in students' mathematical learning so that they are not missing critical and foundational skills needed for the next level of mathematics. The following three examples show very simplistically in three different instances how algebraic reasoning builds from early elementary to high school through vertical alignment.

EXAMPLE 1

One example of extending an elementary math approach into the middle school years is using a chip model. In chapter 5 (addition and subtraction), we examined how you can use chips to model place value. See figure 8.1 for a reminder.

One of the most effective ways to introduce middle school students to adding and subtracting integers is to use counters or colored chips. In chapter 4, we suggested buying counters as a tool for your home. These will come in handy in middle school as students learn to operate with **integers** (positive and negative whole numbers).

Let's consider the equation $5 + (-6) = ?$. This equation can yield a variety of incorrect solutions without proper understanding of the values and operation of the question. Many students will incorrectly answer 1, −11, or even

Hundreds	Tens	Ones	
	OOO OOO	OOOOO OO	⇐ 67
O		OOOOO OOOO	⇐ 109

Figure 8.1.

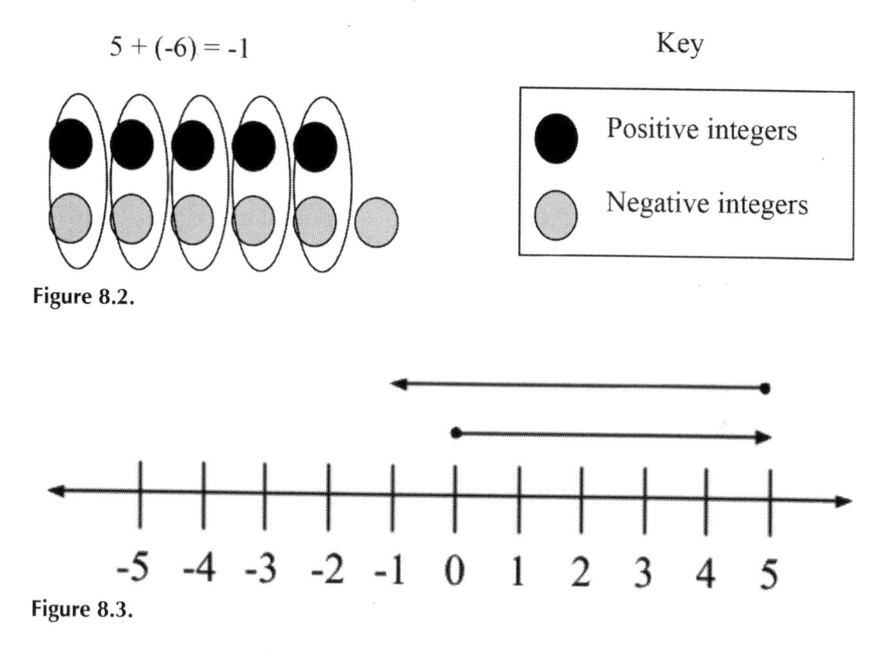

Figure 8.2.

Figure 8.3.

30. Let's use the chips that were explored in the elementary classroom to make sense of this middle school context. The two-colored chips work best for this, as one side can represent positive integers and the other side can represent negative integers (see figure 8.2).

There are five positive integer chips put down. Then, we are adding six negative integer chips to the board. A positive and a negative make a **zero pair** (e.g., $1 + (-1) = 0$). There are five zero pairs identified, so we are left with a single negative integer. This leads to a solution of -1.

Another approach of solving this integer problem would be to use the already established number line. We always start at zero and begin by adding five so we move five places to the right. Next, we are adding negative six, and so we are moving six places to the left. As we get back to zero, we have again formed five zero pairs. We land on -1 on the number line, indicating that $5 + (-6) = -1$ (see figure 8.3).

EXAMPLE 2

In early elementary, students are asked to find an unknown quantity, usually within the context of a story problem. For example, *some* bunnies sat on the grass. *Three more* came over. Now, there are *five*. How many bunnies were sitting on the grass in the beginning? Students might be asked to represent the story with a number sentence and would draw a box or put a question mark (?) to represent the unknown.

$$? + 3 = 5$$

In late elementary school, students are asked to find an unknown quantity represented by a *variable*, or a symbol that replaces a number, usually in the form of a letter.

$$n + 3 = 5$$

Take a look and compare early elementary to late elementary thinking. How does it increase in rigor? Though these two equations may not look that different, the instruction and language teachers use is what makes this a *progression*. Though it doesn't look like much has changed, students in upper elementary school must become adept with inserting the English alphabet into their math! This is quite challenging for students to begin to make sense of!

In middle school, students start to solve one- and two-step equations with an unknown value. For example, students are asked to solve for the unknown x in the problem $3x + 5 = 8$. The foundations for solving for the unknown quantity are based upon the progression of work done in elementary school. First, a student needs to make sense of the number sentence. For instance, "when I add 5 to some number multiplied by 3, I get 8." Understanding equality and that both sides *balance* is critical. Figure 8.4 shows how we represent the equation using an actual balance. Students in middle school should spend a lot of time balancing coins and cubes to recognize how to literally balance equations.

Figure 8.4.

In figure 8.4, we have represented the algebraic equation with three triangles plus five circles on the left and set it equal to eight circles (it's balanced). It's always better to represent this *concretely*, so students should practice actually balancing items before representing it in picture form. How could you figure out what one triangle's value is using the picture? Students recognize that to remain balanced, whatever we do to one side we must do to the other. You can see that we can remove five circles from the left side, which means we need to remove five circles from the right side, too (see figure 8.5).

If we remove the five circles from each side, we will be left with three triangles on the left side and three circles on the right side. Continuing in the same manner to get to one triangle, we can remove two triangles and two circles, leaving one triangle to be equal to one circle. Returning back to the equation, $3x + 5 = 8$, we have now shown that x (or one triangle) equals one.

Compare this approach to the traditional algorithm (see figure 8.6).

What do you notice? It's still important that kids in middle school eventually solve for the unknown quantity using more formalized procedures, but remember, the goal is to understand what they're doing. To do that, we need to start conceptually.

Figure 8.5.

$$3x + 5 = 8$$
$$\underline{\quad -5 \ -5} \text{ (subtract 5 from both sides)}$$
$$3x \quad = 3$$
$$3x = 3 \text{ (divide both sides by 3)}$$
$$x = 1$$

Figure 8.6.

This same idea can be extended to solving for two unknown values in a high school math class. In this instance, instead of just having one variable, x, students would need to find x and y when given two equations. Let's use the conceptual strategy to make sense of high school math using the equations $3x + 2y = 12$ and $4x - 2y = 2$. Let's imagine these equations, where x is represented by a triangle and y is represented by a square. (See figure 8.7.)

Using two balances, we will visualize how to figure out the values of the triangles (x) and squares (y) (see figure 8.8). In figure 8.8, you can see on the left a balance showing that $3x + 2y$ is equal to 12. On the other balance, because $4x - 2y = 2$ has subtraction, representing this equation as written presents a challenge. How does one show taking away two squares on this balance? Instead, let's rewrite the equation so we *can* show it on the balance. Using our knowledge of zero pairs, we can add $2y$ to both sides of the equation to obtain $4x = 2 + 2y$, which is easier to represent on the balance.

If we were to take both balances and redraw them as one balance, it would look like figure 8.9 because we would take the four triangles from the second balance and place them on the left side of the other balance and we would take the two squares and two circles from the second balance and put them on the right side of the other balance.

Now, we have both equations on one balance, and the left side is still equal to the right side. Noticing both sides have squares, we can use the zero pair strategy to eliminate them (see figure 8.10).

$$3\triangle + 2\square = 12\bigcirc$$

$$4\triangle - 2\square = 2\bigcirc$$

Figure 8.7.

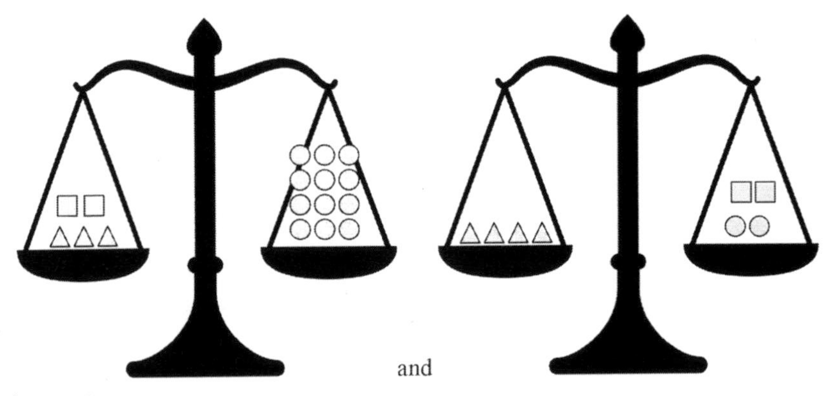

Figure 8.8.

Figure 8.9.

Figure 8.10.

By eliminating the squares, we can easily see now that 7 triangles is in balance with 14 circles. Students can see that each triangle is twice the weight of each circle, meaning that each triangle weighs the same as two circles. If we removed one triangle from the left side, we would need to remove two circles from the right side to keep it balanced. We could continually do this until we are left with one triangle on the left and two circles on the right (see figure 8.11).

The value of one triangle is two circles. Now that we know this, we can return to our original balance of $3x + 2y = 12$ and eliminate 2 circles for every triangle to help us visualize the balance with fewer variables. This means we would eliminate the three triangles on the left side and six circles on the right (see figure 8.12).

Figure 8.11.

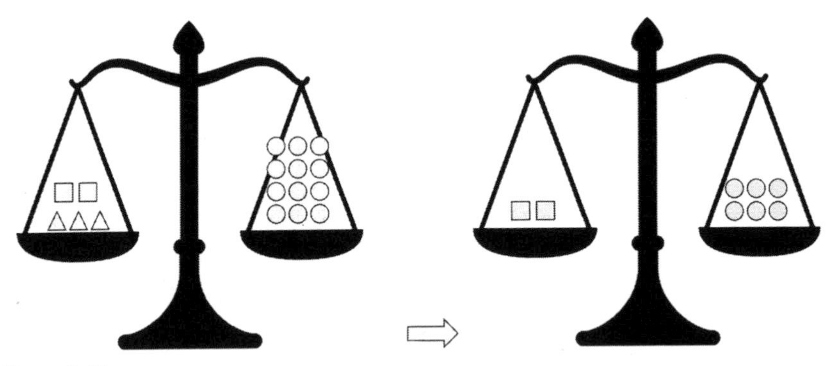

Figure 8.12.

$$\begin{aligned} 3x + 2y &= 12 \\ +\ 4x - 2y &= \ \ 2 \\ \hline 7x \qquad &= 14 \\ x &= 2 \end{aligned}$$

Eliminate one variable (in this case, y) to find the value of x.

$$\begin{aligned} 3(2) + 2y &= 12 \\ 6 + 2y &= 12 \\ 2y &= 6 \\ y &= 3 \end{aligned}$$

Plug the value of x into one of the equations to find the value of y.

Figure 8.13.

It is now clear that for every two squares, there are six circles, which means one square is equal to three circles. Let's return back to our original equations: $3x + 2y = 12$ and $4x - 2y = 2$. We know that $x = 2$ (circles) and $y = 3$ (circles). Let's plug those values in to see if the equations are true. $3(2) + 2(3) = 12$ and $4(2) - 2(3) = 2$. It works! $6 + 6 = 12$ and $8 - 6 = 2$.

Once students have had ample time applying their thinking from elementary and middle school to high school, they then should learn the procedures to find efficient ways of communicating the mathematics (in this case, the elimination procedure that is commonly used in solving this type of problem). (See figure 8.13.)

In this way, students first experience solving simultaneous equations with meaning and then apply the meaning to the process.

EXAMPLE 3

In chapter 6, we explored how the area model is used to understand multiplication and division. Let's look back at one of the multiplication problems, 32×25, to see how the area model will be used beyond elementary school (see figure 8.14).

Now imagine we are in a middle school classroom. The children are asked to find the value of $(3x + 2)(2x + 5)$. This is almost identical to the problem of $(30 + 2)(20 + 5)$. Instead of a tens value being added to a ones value, it is a variable being added to a ones value, but the idea is exactly the same. Kids who had a strong conceptual understanding of multiplication of whole numbers using the area model will be able to apply their knowledge to this new

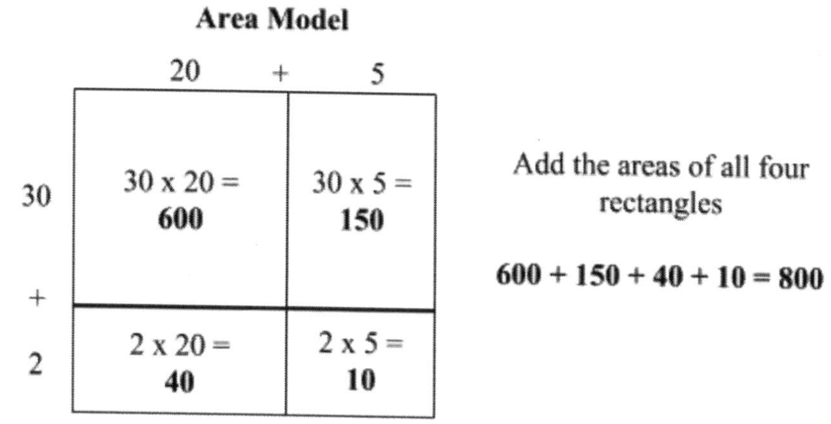

Area Model

	20	+ 5
30	30 x 20 = **600**	30 x 5 = **150**
+ 2	2 x 20 = **40**	2 x 5 = **10**

Add the areas of all four rectangles

600 + 150 + 40 + 10 = 800

Figure 8.14.

problem-type, so long as they understand that x is a variable that represents a quantity. Let's look at the same area model, but this time, show what it would look like for $(3x + 2)(2x + 5)$. (See figure 8.15.)

Take a moment and compare figures 8.14 and 8.15. Can you see how incredibly similar the two problems are? In fact, if you make $x = 10$ in figure 8.15, then you end up with exact same model and solution as the problem in figure 8.14!

This idea can be extended further into high school mathematics for the product of expressions involving many terms. Very easily students can multiply and combine like terms. Expanding on the previous example, we add x^2 to make the first factor a *trinomial* (an algebraic expression consisting of three terms) and the process does not change. We can find the product of $x^2 + 3x + 2$ and $2x + 5$ just as we have with all other area models by finding the areas of smaller rectangles inside of the large one (see figure 8.16).

With a solid understanding of an area model, students can easily multiply trinomials by trinomials with a perfect understanding of what they are doing and why they are doing it. This approach is much better than the way we as parents learned distribution using FOIL, a procedure for distribution. FOIL stands for First, Outside, Inside, Last (see figure 8.17), and we strongly advise against teaching it, as it does not offer conceptual understanding, nor does distribution have to be done in the order in which it suggests. When multiplying *binomials* (algebraic expressions with two terms), it works like a charm; however, when students try to take this idea of FOILing and apply to multiplying trinomials, it will no longer work as the middle term is inadvertently ignored, and not distributed.

This approach is still sometimes taught in Algebra I courses, but it is ill advised. We like to say there is no such thing as FOIL in mathematics.

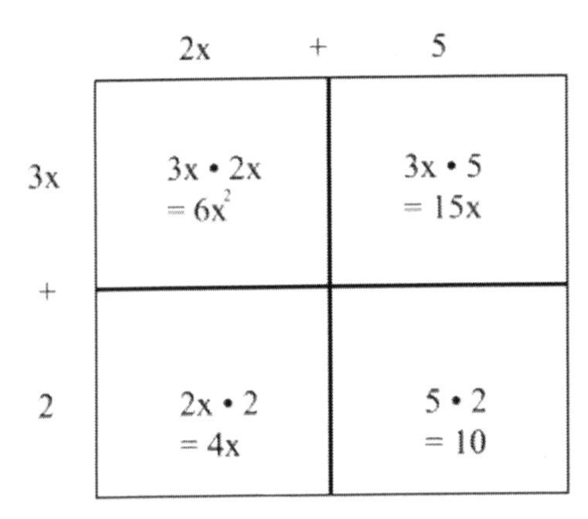

Add the areas of all four rectangles

$6x^2 + 15x + 4x + 10$

Figure 8.15.

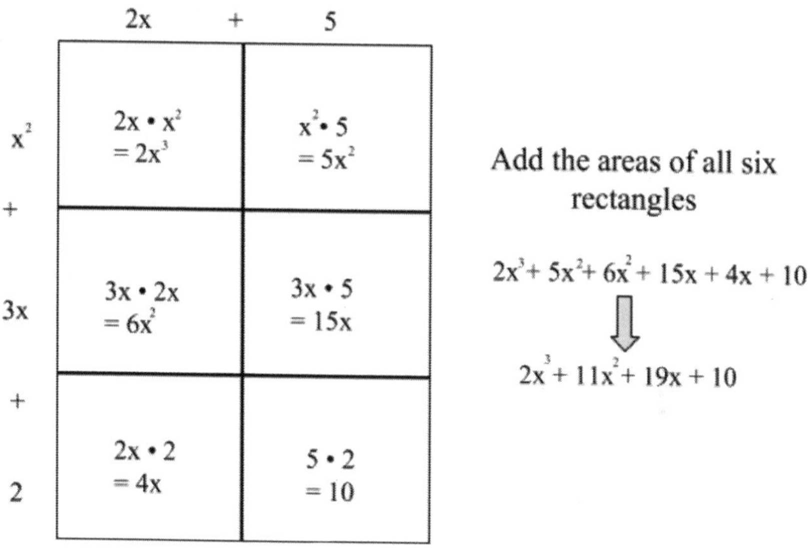

	$2x$ +	5
x^2	$2x \cdot x^2 = 2x^3$	$x^2 \cdot 5 = 5x^2$
+ $3x$	$3x \cdot 2x = 6x^2$	$3x \cdot 5 = 15x$
+ 2	$2x \cdot 2 = 4x$	$5 \cdot 2 = 10$

Add the areas of all six rectangles

$$2x^3 + 5x^2 + 6x^2 + 15x + 4x + 10$$

$$\Downarrow$$

$$2x^3 + 11x^2 + 19x + 10$$

Figure 8.16.

First Terms → $(2x+3)(x-2) = 2x(x) = 2x^2$

Outer Terms → $(2x+3)(x-2) = 2x(-2) = -4x$

Inside Terms → $(2x+3)(x-2) = 3(x) = 3x$

Last Terms → $(2x+3)(x-2) = 3(-2) = -6$

This results in:

$2x^2 - 4x + 3x - 6$

Combine like terms: $(-4x + 3x = -x)$

$2x^2 - x - 6$ This is the final answer.

Figure 8.17.

Instead, the single idea of the area model can be taken from third grade through high school.

As you can see, learning math conceptually (for understanding) before procedurally (step-by-step process) will provide students a solid foundation in each domain of mathematics and allow them to use prior knowledge to help them apply what they've learned to new problems in the future. Overall, as your child progresses from upper elementary to middle school, we encourage you to focus on the following three main areas to ensure your child is ready for the more complicated applications of middle and high school mathematics.

NUMERICAL FLUENCY

Numerical fluency is a strong sense of numbers that allows students to understand how numbers fit together and how numbers can be grouped in order to do mental computation. For example, there are a variety of ways that the problem 24×4 can easily be solved mentally with strong numerical fluency. Using decomposition, one might break apart 24 into 20 and 4 and use partial products $(20 \times 4 + 4 \times 4)$ to identify 96. Alternatively, one could think of 25×4, a fact most know by heart, and then subtract the extra one group of four that was rounded $(25 \times 4 - 1 \times 4)$. Another possible option is to use a doubling and halving strategy by dividing 24 by 2 and multiplying 4 by 2 to change the problem to 12×8, a fact that might be more manageable to think about.

PROFICIENCY WITH FRACTIONS

There is a strong correlation between a student's ability to perform operations with and understand the concept of fractions and their success in algebra. Spending time to understand fractions in a conceptual manner will lead to flexible thinking when studying algebra. Spend time at home identifying fractions around you (e.g., *three-fifths* of the doors in our home are wood), cooking and baking with fractional measurements for real-world context, or practicing counting by fractional amounts just as you do with whole numbers (e.g., *one-eighth*, *two-eighths*, *three-eighths*, etc.).

ALGEBRAIC REASONING

Algebraic reasoning is the heart and soul of mathematics. It is the ability to explore mathematical structure and recognize relationships and patterns between quantities. Being able to reason algebraically means thinking beyond calculations and arithmetic and focusing more on functional thinking, relationships, and patterns. Being able to generalize rules from repeated instances is important to build in elementary school, as well as in middle school. You can encourage your child to build algebraic reasoning skills by asking them to justify and prove their thinking and to predict in math.

As your child becomes proficient with these topics, they will be able to understand proportional relationships and find unknown values such as length, angles, areas, and perimeters using algebraic equations, formulas, and graphs. Algebra requires the use of all these topics and is pivotal in students understanding the high school standards. Students will be able to use their foundational algebra knowledge to perform advanced calculations such as finding mortgage and car payment amounts, compare compound interest rates, and understand how credit companies calculate interest and payments to determine financing costs.

Chapter 9

What You Can Do at Home

"Parents don't really think about math as something you do at home, they think about it as something you do in school."

—Deborah Stipek, Stanford University

So, the big question for parents becomes "How can I best help my child with math?" There are two avenues in which you can help. Avenue one is with the math itself, and avenue two is with fostering the positive habits of a mathematical thinker. As parents, it is easy to forget the woes of school, and for some, math can be particularly confusing, frustrating, and difficult, or be a reminder of past struggles. As a parent, it is also difficult to know how to best help your child. Parents tend to either help their child too much because they want them to get good grades and don't want to see them fail or struggle at something, *or* they don't help their child enough because they didn't have a good experience with math themselves and feel unqualified to help them.

Let's start first by setting the scene and creating an atmosphere of learning. First, it is important to have high expectations for your child. When you believe your children can learn math, they will rise to the occasion—so set the bar high! Believe that they can do it and then help them to discover the resources they need in order to be successful. Let's look at some ideas for how you can create and engineer positive "learning experiences" at home, and how that evolves through ages and stages of development.

The following list of activities and materials are not rigid; you may find that your child is ready for some of the suggestions listed in the higher grade levels, or that they might enjoy and benefit from some of the suggestions from an earlier stage. This is totally fine! The main goal is for you to find ways to make learning about mathematical concepts easy, stress free, and enjoyable for your child(ren). Peruse each section and pick and choose activities that you can use at home.

PRESCHOOL–KINDERGARTEN

Games

Playing games with your children can be formal or informal. Children love to play games, and math and play are often not connected enough. Encourage your child to try the following:

- Count blocks as they build wondrous towers.
- Ask them to sort a group of toys and see if they sort by size, color, or shape. Once they are finished ask them if they could sort them in a different way. Ask them how they chose their method of sorting and if they can think of a different way to sort.
- Use Monopoly money to play "store."

The following traditional card and board games have many opportunities to develop the foundation for mathematical concepts:

Go Fish. Playing cards are dealt out to each player, and you take turns asking for cards trying to make matches. A variation on this can occur where players ask for cards to make a sum of 10. For example, if you have a seven, you would ask another player to see if they have a three to earn a point.

Memory. Using playing cards or special cards put facedown, players select two cards to try and make a match. They need to remember where cards lay after they are turned over. There are some great problem-solving strategies to be developed when playing.

Chutes and Ladders. This game is a dice-rolling game that requires students to count and move from 1 to 100. Landing on certain spaces will advance or regress the progress. In addition to basic counting, the introduction of probability can be explored with this game.

Trouble. This board game uses a popping die to determine the number of moves for each piece, but there is mathematical strategy that can be discussed.

Tiny Polka Dot. This box includes 16 easy-to-learn games that teach math in a fun way, from counting and early numeracy to arithmetic and logic! This game builds problem solving and number sense.

Books and Literature

Read books that have mathematical themes. A great list of books for this stage:

- *7 Ate 9*, by Tara Lazar
- *Goodnight, Numbers*, by Danica McKellar
- *Chicka Chicka 1, 2, 3*, by Bill Martin Jr.
- *Ten Magic Butterflies*, by Danica McKellar
- *One Hundred Hungry Ants*, by Elinor J. Pinczes

FIRST–SECOND GRADE

Games

- Discuss and use math with your child while cooking.
- Partition cookies or pizza evenly and then decide how to divide the remainder equally among the family (wonderful for fractions).
- Double, triple, or quadruple serving sizes, encouraging them to add repeatedly.
- Count the days on a calendar. You can post a calendar in your home and use it on a daily basis or count how many days are left in a month or until an important event.
- Use your feet, paper, and other nonstandard tools for measurement to find the length and width of a room, table, or picture. Estimate the dimensions first and then measure to see how accurate you are. For an extension, use standard tools for measurement, such as a ruler or tape measure, to measure.
- Play "I Spy" to find different shaped items within the house, and encourage your child to identify the shape and explain how they know it's the shape they identified.
- Do math while you're driving. Interstate highways work perfectly for this. In many U.S. states, the exit numbers indicate how many miles you are from the state border. Do some addition and subtraction! You can even identify numbers as odd or even as you drive up and down streets with houses. Some streets only have odds on one side and evens on the other side—let your child come to this generalization! Noticing speed limits or license plates and talking about the numbers is another great method.
- Have a supply box at home for your child. This can be turned into a fun activity in which you allow your child to decorate a shoe box and fill it with the supplies that they might need to complete any assignment including pencils, pencil sharpener, pens, compass, protractor, scrap paper, crayons,

glue, scissors, a ruler, paper, note cards, highlighters, and paper. It is also a good place to store flash cards that you can either purchase or have your child make. We recommend that you have your child make their own so they develop a useful study strategy.

The following traditional card and board games have many opportunities to develop the foundation for mathematical concepts:

Monopoly Jr. Engage with mathematics in a real-world situation by using play money to rent and purchase real estate. Have your child be the banker so they are practicing addition and subtraction and making change!

Chess. Chess is a game of strategy that teaches you to think ahead of your opponent. The game is excellent for building problem-solving skills.

Checkers. Similar to Chess, Checkers is a game of strategy and forces children to think ahead to how their choices and actions could impact the future of the game. The game enhances problem-solving skills.

Sum Swamp. Practice addition with this fun game! This game is a great way to build early numeracy and mental math skills.

Books and Literature

A great list of books with mathematical themes for this stage include

- *Bedtime Math: A Fun Excuse to Stay Up Late*, by Laura Overdeck
- *12 Ways to Get to 11*, by Eve Merriam
- *17 Kings and 42 Elephants*, by Margaret Mahy Alexander
- *Alexander, Who Used to Be Rich Last Sunday*, by Judith Viorst
- *Amanda Bean's Amazing Dream*, by Cindy Neuschwander

THIRD–FIFTH GRADE

Collaboratively, create a space and routine with your child where they feel successful studying and doing homework. Allowing a child to work in their room where they can become distracted by electronics and games may not be a good idea; however, for others, they can become easily distracted if they are in a central location within the home, such as the kitchen island or table.

Spending time to problem solve this with your child will help them learn about their own preferences and help them advocate for the variables that they need to create a productive learning environment for themselves. As they grow and develop, their work space may change with them. Be supportive of this, and also make yourself available to provide assistance, encouragement, or to guide them in finding the resources they may need to complete assignments.

Games

Family game nights are a great way for families to encourage their children to enjoy mathematics. There are numerous games that have a strong mathematical foundation, while also being enjoyable for kids and adults. We list some below.

Albert's Insomnia. A beautiful, yet simple card game that helps authentically develop the basic operations. Players can work collaboratively or competitively to determine different values using two, three, or four digits.

Prime Climb. This brilliant game focuses on building strategy with prime numbers. Players build up to the number 101 using dice and mathematics. Bonus cards are given when your turn ends on a prime number.

24 Game. This creative game is similar to Albert's Insomnia with a focus on basic operations; each player is given four cards and needs to use all four exactly one time to end with the number 24. There are varying degrees of difficulty, but kids of all ability groups love to play the game.

Head Full of Numbers. Shake the "head" cup to roll the dice and, in a given amount of time, write as many equations as you can from the numbers and symbols on the dice before the time is up! We suggest you remove the timing aspect to promote learning, not math anxiety.

Pay Day. Practice finance at home through this fun game where you earn money and then spend it! This game reinforces decision making and analyzing choices.

Mathopoly. A math take on the classic Monopoly game.

Technology

Advocate the use of technology judiciously. The internet is a wonderful resource, if used properly. There are copious amounts of math tutorials, videos, and practice problems for free. However, there are also some sites that allow students to take a picture of a problem and it will give them an answer.

It is important that we teach children at a young age that it is not about the answer, it is about the process. Getting a correct answer does not always equate to understanding how they got the answer. The goal is for students to be able to apply skills from one type of problem to another. An answer should be nothing more than proof that there is understanding of the process. (This is why teachers insist on students showing their work particularly for skills that they know are critical for a different, more complex problem.) Some popular online games and applications include

IXL. https://www.ixl.com/math/
Dreambox Learning. www.dreambox.com/
ALEKS. https://www.aleks.com/independent/parents
Splash Math. https://www.splashmath.com/features/parents
Greg Tang Math. http://gregtangmath.com/games

Books and Literature

A great list of books with mathematical themes for this stage include

- *The Grapes of Math*, by Greg Tang
- *How Many Guinea Pigs Can Fit on a Plane?* by Laura Overdeck
- *Hidden Figures*, by Margot Shetterly
- *The Number Devil: A Mathematical Adventure*, by Hans Magnus Enzensberger
- *A Place for Zero*, by Angeline Sparagna LoPresti

FIFTH GRADE AND BEYOND

It is important to model how to effectively learn and study. Require that your child have notes from the class, a textbook, or whatever information the teacher has given the student during class time right next to their homework paper. Before answering any questions, have them find a similar problem in their materials. Ask them to look at it first and see if they can figure out what they need to do.

Never give the "answer" to the problem! This keeps them from developing the skills they need to take responsibility for their own learning. Pretty soon you are on a merry-go-round where you child will not attempt the work and wait for you, or someone else, to provide the answer. It is difficult watching a child struggle, but remember that it is our job to prepare children for adulthood where struggling is pretty much a fact of life. It can also be misleading to a teacher and create a situation in which a student is not receiving the help they need in order to master the concept. After the test is too late to discover that a student doesn't understand the concept and the processes involved.

It's also critical that you teach your child how to ask a friend for help. If they are stumped or cannot get started, they should ask, "What is the first thing I should do?" Then they should try to complete the problem from there. If they need additional help, it should include questions such as "Why do I make this step?" The only time they should ever ask a friend for an answer is when they are verifying that they themselves have the right answer. If they have different answers, then encourage them both to rework the problem and see where they worked the process differently or where they made a computational error.

CONSIDERATIONS FOR ALL AGES

Always talk about math in positive ways. Never tell them that math is too hard and that you couldn't do it either. This creates an excuse for not trying. It is critical that you let them know that math is very important and always encourage them to do their best. "Never give up" should be the mantra.

Ask questions that lead to self-discovery. Here are some examples:

- How is this problem different from the ones you have already worked?
- What do you know about this already?
- What is usually the first step?
- Can you find a problem similar to it in your notes or one that you have already worked?
- Is your answer reasonable?
- What resources have you already looked at that might help you understand what you need to do?
- If we have to look up a tutorial video, what would we search for in order to find the right video?
- Have you checked to make sure that you haven't made a simple computational error?
- Have you read the directions carefully and understand what you are being asked to do?

If necessary, learn as your child learns. The internet has opened up a whole new world where anyone of any age can learn a new skill. There are a copious amount of math tutorials, videos, and practice worksheets for free.

After your child has completed an assignment, ask them to "teach" you how to do it. Ask them "why" and "what if" as often as possible. This allows them to take their learning one step further into an in-depth understanding of a process.

Make math an everyday part of your life. Spend time with your child playing any type of game that involves math, or, for the really creative, have them make up a game that you can play together. For example,

- Show them how you use math in either your job or your home life. Notice the math around you.

- Point out geometric shapes in the real world or ask them to find shapes for you and justify how they know they are those shapes.
- Engage your child in conversations about math in the real world such as, "If I put 10 gallons of gas in my car and I get 20 miles per gallon, how many miles can I drive?"
- Look for parallel and perpendicular lines around you.
- Have them use a calculator to calculate the cost of groceries and then find the tax amount and total cost.
- Have them estimate the cost of a small number of items. For example, if snow cones cost $3.19 each, about how much money will I need in order to buy the whole family a snow cone?

The more you look the more you will see math in the world around you, and so will they.

Last, but not least, be a proactive part of your child's education. You can do this by trying the following:

- Attend parent teacher conferences with an open mind as to how you can best help your child. It feels really good when we only hear praises about our children, but it is critical that you understand areas of improvement needed so that you can help them. Ask the teacher for specific ways in order to help with mathematics.
- Be familiar with the school, the website, and any other digital means of tracking your child's progress. Most schools now have an online gradebook in which you can see how your child is performing on assessments. It is much easier to hone in on one concept in which they are struggling than it is to get to the end of the nine weeks and find out they are now missing multiple skills.
- Send your child to school ready to learn. They must have a good night's sleep and good breakfast, the needed supplies (pencils, backpack, etc.), and a positive attitude toward learning.
- Attendance is crucial. Your child cannot learn when they are not present to learn. When a student has a chronic absentee problem, they are in a perpetual state of playing catch-up. They can't focus on new concepts because they haven't mastered prerequisite skills.

DEVELOP POSITIVE HABITS OF MIND

Teaching your elementary-aged child problem-solving skills and typical habits of mind of math thinkers will have big payoffs at the middle school level and then, later, into the high school level. Chances are they will approach a point in their schooling where the mathematics is beyond your ability to help them without taking a refresher course yourself. Ultimately, your goal should be to teach them how to problem solve independent of you. This will prove to be useful not only in mathematics but also most likely will apply to all other aspects of their lives.

There are a few habits of thinking that we encourage parents to try to build within their child at home.

Productive Disposition

When people have a productive disposition, that means that they see value in what they are doing—they recognize that what they are learning is worthwhile. Kids who have a productive disposition see themselves as mathematical thinkers and understand that their mindset is important to help them get better at the subject. To help your child develop a productive disposition at home, encourage them to become curious thinkers. Praise them when they question anything and everything around them and help them see that you, too, believe you are capable at doing math.

Perseverance

Children are often so quick to seek out help when they struggle with a problem. Rarely will they spend more than 30 seconds before asking for the answer or a hint. Knowing problem-solving strategies that can be used in any situation can help everyone with perseverance. They include using guess and check, creating a picture or diagram, working backward, making a table, and looking for patterns. Having many different strategies to approach a difficult problem will promote perseverance.

Justification of Ideas

Having students write and articulate their ideas not only helps build their understanding of the concept but also improves their thinking and reasoning skills. Often, students are asked to explain or justify their solutions, but they have never been instructed how to do this. Justifying means that you are able to explain the steps that you took and why you took them. Help your child be able to articulate why they know that their answer is correct and always ask them to show more than one way for how they came to their answer.

Evaluation of Others' Ideas

How often are children presented with a problem, solution, and steps that someone else took to arrive there? Not very often, and this is a great activity to do with your children. Share with them a common problem and a different method for solving it. Have them work to understand the process that you used and ask them to evaluate whether one process was more efficient than another.

Connection of Math to the Real World

Often textbooks try and force the real-world component of math with examples that are borderline silly. However, the world we live in is surrounded with beautiful examples of mathematics. Help them see math wherever they go! If you're in a bathroom, look at the floor or ceiling tiles and see if you notice a pattern or see symmetry. If at the soccer fields, have them estimate the distance between the two goals. If in the car traveling a certain speed, have them determine how long it will take to travel to your destination.

Use of Resources

Encouraging your child to identify and use appropriate resources and tools is an important skill of a mathematical thinker and also helps in developing independence. Spend time working with your child to learn the different functions of a calculator or how to use a ruler or protractor. Urge your child to use things around them as tools if the actual tool they need is not immediately available (e.g., if they need a ruler, suggest sticky notes, or using their feet, to estimate).

Precise Communication

Being precise is an art; there are times when being accurate and exact matters, such as in tiling a bathroom shower, but there are other times when it's not as needed, such as estimating. Determining the different situations in which an exact answer is needed and when an estimate will do is a great activity for your children.

Make Conjectures

Mathematical thinkers predict and theorize based on repeated experiences of the past. Help your child make their own theories, or conjectures, by asking them, "what do you think would come next?" or "what would happen if . . . " After, help them determine a method to be able to test their conjectures.

Bottom line? Make math a part of your everyday life at home. This will foster a positive mindset for your child and help them see that math should be fun, not scary or impossible.

Chapter 10

Conclusion

> Mathematics is not about numbers, equations, computations, or algorithms: it is about understanding.
>
> —William Paul Thurston

We applaud you for taking an active role in helping your child succeed in mathematics. We realize this book does not answer every question you might have, but we hope it gives you a fresh look inside of your child's elementary math class and helps you see a new side to the way math is being taught today. The elementary years are often the make or break time for students to develop a passion for math, and the support you provide is crucial. We conclude the book with a top 10 list of how best to support your child.

1. HAVE A GROWTH MINDSET

It is important for everyone in the family to showcase a positive and open mind with regard to math. There is no such thing as a "math person," and your ability (or perceived inability) to do mathematics is not genetic. Watch your language around your kids and stop yourself before you say things like, "I wasn't good at math" or "Ask your father; he's the math person in this family."

2. HELP BUILD UNDERSTANDING

Resist the urge to immediately show your child the quick process for doing a computation. Take time to understand how the teacher and curriculum are approaching the skill and try to support that process. Always revert back to the Concrete-Representational-Abstract (CRA) approach to ensure your child is learning an idea conceptually.

3. ASK QUESTIONS INSTEAD OF TELLING

It is important for kids to struggle in a productive manner. To assist when appropriate, we should ask questions of them to elicit more thinking. Your ultimate goal is to enable them to be independent of you. Whenever they ask you a question, be like a teacher and ask a question back rather than tell them an answer or statement.

4. INCLUDE DAILY MATH TALK AT HOME

Math is joyful, and math is everywhere. Recreate this message in your daily lives by talking mathematically at home. This is more than a "How was your math class today?" sort of conversation. Instead, it's about engaging in interesting conversations.

Ask your child things such as, "Would you rather receive a penny doubled each day or $10 a month for your allowance? Why?" or "About how many shirts do you predict are in this laundry basket? Let's fold them and find out." The Table Talk Math Mat is a great tool to assist in daily math conversations at the dinner table. You can purchase it at https://squareup.com/store/tabletalkmath.

5. NOTICE MATH AROUND YOU

As stated previously, math is everywhere. Make a concerted effort to point out things you see around you that are mathematical. For example, at the grocery store, ask your child to predict the weight of a bag of apples before you purchase them. Let your child use the scale and weigh them to see how much they weigh. Show your child that the double yellow lines on a street are *parallel* (lines that will never intersect). Point out palindromic numbers on the digital clock in the car (e.g., 12:21; 10:01; etc.).

6. HAVE YOUR CHILD DO MATH BEFORE BED

Most parents encourage their children to read before bed, but rarely do they say, "Be sure to do some math before bed!" We highly encourage integrating a bedtime math routine much like a reading routine. Even if it's just five minutes of playing Sudoku, KenKen, or some other math puzzle, your child will benefit. Two apps we suggest to foster this routine are http://bedtimemath.org/apps/ and https://mathbeforebed.com/.

7. PRAISE EFFORT

Kids love to hear that they've done something well, and it's important when we praise or compliment children that we focus on their *efforts* and *behaviors*, not on their intelligence. Resist the urge to call your child a "genius" or "smart." This creates the belief that intelligence is innate.

Instead, focus on the behaviors that produce the things you want to praise. For example, praise a child's perseverance when a task is challenging instead of making a big deal about the right answer. This will implicitly tell your child that you always want to see them work through a challenging task, not that you always want to see a correct answer.

8. USE THE INTERNET AND ONLINE RESOURCES

We live in an incredible era where information is at the tip of our fingers. If your child comes home with homework and there is some math that is not referenced in this book, be proactive and search the internet for it yourself. Tape diagrams and bar models, for example, are problem-solving tools that we did not cover, yet you might see your child using them. A simple internet search will populate hundreds of video tutorials, worksheets, and readings that will help you better understand the tool.

9. CREATE A MATH TOOLBOX AT HOME

In order for your child to become more independent and rely less on you, a math toolbox or supply center in their workspace is one of the best ways to achieve this. By putting rulers, markers, crayons, string, tape, calculators, graph paper, and more in an accessible place, your child no longer needs to ask you for the tools, as they are within arm's reach.

10. COMMUNICATE WITH YOUR CHILD'S TEACHER

One of the most informative tools is your child's teacher. Be sure you are in contact with them *before* an issue occurs. Start the school year off introducing yourself and letting them know that you are excited to learn more about how kids are learning math. Let them know you have read this book and that you might ask questions so that you can be most helpful. Teachers love when parents communicate positively and proactively. It makes both roles easier.

Though the book has come to an end, your role as a learner is just beginning. Each and every day, mathematical thinkers learn something new or apply something they've learned in the past in a new way. What changes are you going to make as a result of finishing this book? What are your goals for parenting an elementary-aged child who's learning math? How will you reinforce that math is fun and joyful and all around you? How will you spread the word and inform other parents?

Glossary of Terms

It is our goal with this book to make elementary math as accessible as possible for parents. We realize that, in doing so, we may use educational phrases that are not as familiar to you. We encourage you to take some time to read the words and become familiar with them since they will be referenced in the chapters of this book. As you read, be sure to come back to the glossary to clarify your thinking or remind yourself what certain words mean. In addition, there are probably thousands of words we could have included, but we selected only the ones that we felt were pertinent to the understanding of the chapters. You will hear your child say most of these words at some point in their elementary math career. We hope this glossary helps you feel as though you are speaking the same language as your child.

abstract. In the CRA approach, A stands for abstract, or the symbolic stage. The expression $3 + 4 = 7$ is abstract because it is made up of symbols with no context.

addend. Addends are the parts of an addition sentence. In the equation $3 + 4 = 7$, 3 and 4 are the addends.

area models. An area model is a visual representation in which the length and width are configured using either multiplication to figure out the area of a larger rectangle or division to figure out the length given an area and a width. This area model in figure G.1 shows $22 \times 5 = 110$.

array. An arrangement of objects or pictures in rows and columns. Figure G.2 is a two-by-four array.

bar models. Bar models are visual models that use rectangles to represent parts and wholes. Also known as tape diagrams, these drawings illustrate number relationships and can sometimes be used to show comparisons. (See figure G.3.)

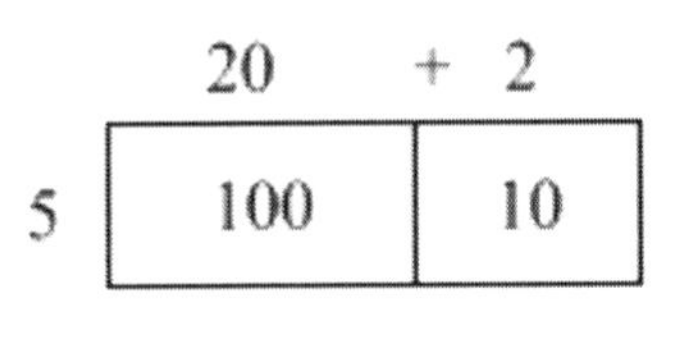

Figure G1.

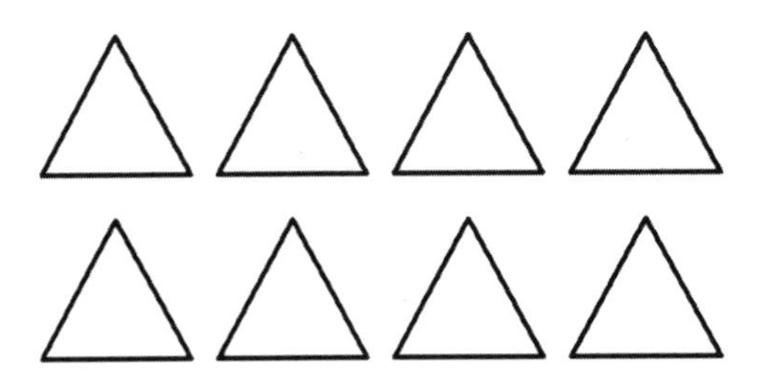

Figure G2.

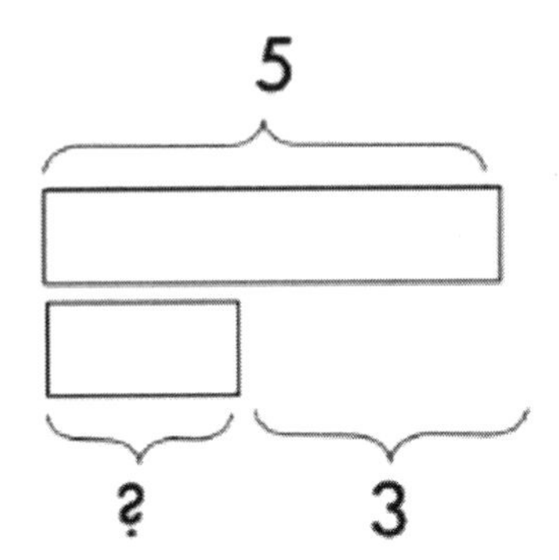

Figure G3.

base ten. Base ten refers to the numbering system in common use that uses decimal numbers. Base ten is also called the decimal system. In base ten, each digit in a position of a number can have an integer value ranging from 0 to 9 (10 possibilities): $\frac{1}{10}$, $\frac{1}{100}$, 1, 10, 100, 10,000, 100,000, etc.

benchmarks. A point of reference that can be used to compare things—you might hear your child refer to "landmark numbers" or "friendly numbers," which are benchmarks (e.g., when working with fractions, $\frac{1}{2}$ is a benchmark, or reference point, used to compare fractions).

cardinality. The notion that the last number said in a counting sequence represents the total objects in a set (e.g., 1, 2, 3, 4—four is the number in the set).

cardinal numbers. Numbers that represent quantities (e.g., one, two, three), as opposed to ordinal numbers, which represent a particular order (e.g., first, second, third).

compose. To put together, to be made up of (e.g., 241 is composed of 200 + 40 + 1; a regular hexagon is composed of six equilateral triangles).

conceptual understanding. A deep comprehension of mathematical ideas that goes beyond knowing procedures and/or facts (e.g., using the properties of numbers to make a problem easier to solve rather than using an algorithm without understanding).

concrete. This stage is the "doing" stage, using hands-on objects to model problems (e.g., using base-ten blocks to add or subtract numbers).

decompose. To break apart (e.g., 32 can be decomposed into 30 + 2 or 10 + 22, etc.; a regular hexagon can be decomposed into six equilateral triangles).

denominator. The bottom number of a fraction (e.g., in the fraction $\frac{a}{b}$, b is the denominator).

difference. The result of subtracting one number from another; how much one number differs from another (e.g., in the equation $7 - 4 = 3$, 3 is the difference).

dividend. The number to be divided (e.g., 384 is the dividend in the expression $384 \div 4$).

divisor. The number being divided by (e.g., 4 is the divisor in the expression $384 \div 4$).

doubles facts. Addition expressions where both addends are the same number (e.g., 6 + 6; 8 + 8; 12 + 12, etc.).

equivalent fractions. Fractions that share the same value but have different numbers of pieces to the same whole (e.g., $\frac{1}{2}$ is equivalent to $\frac{3}{6}$).

expanded form. A multi-digit number written as the sum of the values of each digit (e.g., $321 = 300 + 20 + 1$).

face. In geometry, a face is the flat surface of a solid (e.g., the cube in figure G.4 has six faces).

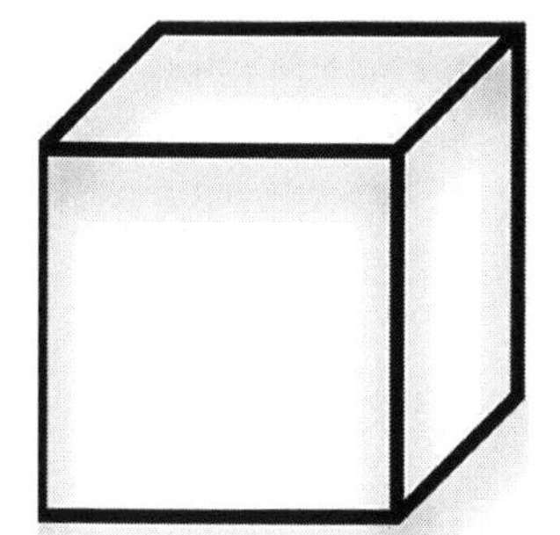

Figure G4.

factor. A whole number or quantity that when multiplied with another whole number produces a given number or expression (e.g., 4 and 12 are factors in the expression $4 \times 12 = 48$).

integers. Positive and negative numbers not written as fractions (e.g., −3, −2, −1, 0, 1, 2, 3, etc.).

kinesthetic. Tactile learning in which students learn through physical movement (e.g., students lining up in a classroom to represent points on a number line).

manipulatives. Objects designed so that a learner can perceive some mathematical concept by manipulating them, hence the name. The use of manipulatives provides a way for children to learn concepts through developmentally appropriate hands-on experience (e.g., base-ten blocks, tiles, pattern blocks, cubes, spinners, etc.).

minuend. The quantity from which another number is to be subtracted (e.g., in the equation $4 - 3 = 1$, 4 is the minuend).

multiples. The result of multiplying a number by a positive or negative whole number (e.g., $3 \times 4 = 12$; so, 12 is a multiple of 3 and 4).

multiplicand. A number that is to be multiplied by another, also known as factor (e.g., in the expression 3×4, 4 is the multiplicand).

multiplier. The number to be multiplied, also known as factor (e.g., in the expression 3×4, 3 is the multiplier).

nonstandard measurements. Units that aren't typically used to measure objects and aren't associated within each measurement system (e.g., paperclips, pencils, or a thumb can act as measurement tools).

nonunit fraction. Fractions that are not written with a numerator of 1 (e.g., $\frac{3}{4}$, $\frac{5}{2}$, and $\frac{2}{3}$ are all nonunit fractions).

number bond. A representational model that shows a quantity or numeral broken down into its parts and specifically showing the relationship between the parts and their whole (e.g., the two parts [3 and 7] make the whole [10], as shown in figure G.5).

number line. A line on which each point represents a real number. It is a geometric representation of numerical values. Some number lines are called "open" or "empty" because students are the ones who decide what values to place rather than it being given to them. (See figure G.6.)

10	
3	7

Figure G.5.

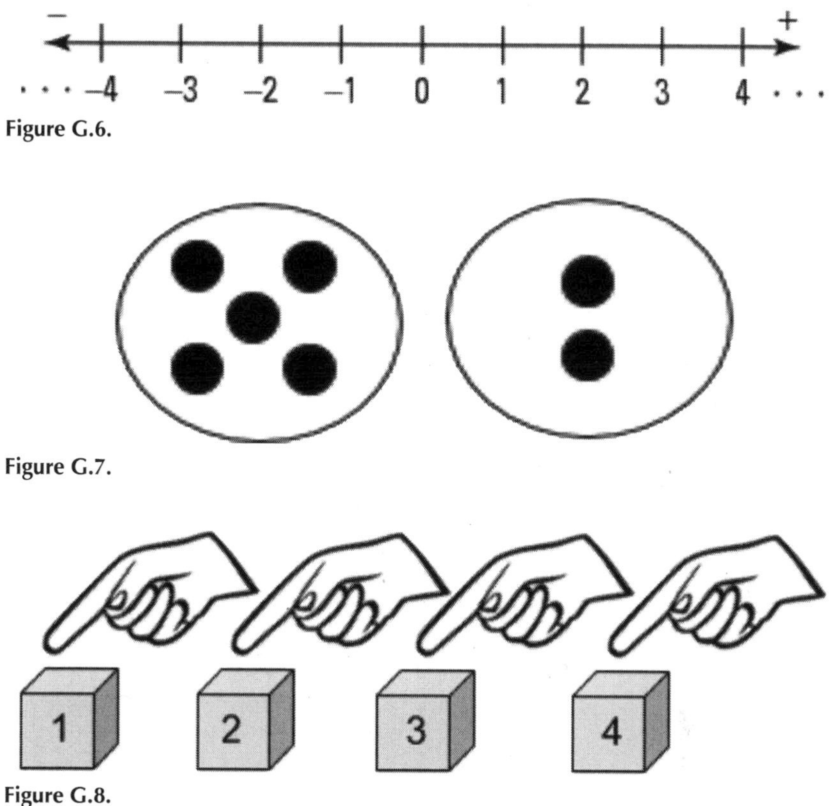

Figure G.6.

Figure G.7.

Figure G.8.

numerator. The top number of a fraction (e.g., in the fraction $\frac{a}{b}$, *a* is the numerator).

numerosity. The innate ability to recognize that two quantities may be different (e.g., in figure G.7, the group on the left has more dots than the group on the right).

operations. Calculations (e.g., ×, ÷, +, and – are operations).

one-to-one correspondence. The ability to connect one number word to one quantity through counting. (See figure G.8.)

partial difference. The subtraction of parts by place value, rather than using the traditional algorithm. With partial differences, students work on one place value at a time (in any order), and then add all the partial differences to find the total difference (e.g., 70 – 20 = 50, and 3 – 5 = −2, so 50 – 2 = 48). (See figure G.9.)

partial product. The product of parts (usually parts of different place value) of each factor; a part of a final product that is the result of multiplying the ones, tens, hundreds, and so on. (See figure G.10.)

$$\begin{array}{rcr} 73 & \rightarrow & 70 + 3 \\ \underline{-\ 25} & \rightarrow & -\ \underline{20 + 5} \\ 48 & \rightarrow & 50 - 2 \end{array}$$

Figure G.9.

$$\begin{array}{r l} 25 & \\ \underline{\times\ 15} & \\ 25 & = 5 \times 5 \\ 100 & = 5 \times 20 \\ 50 & = 10 \times 5 \\ \underline{+\ 200} & = 10 \times 20 \\ 375 & \end{array}$$

Figure G.10.

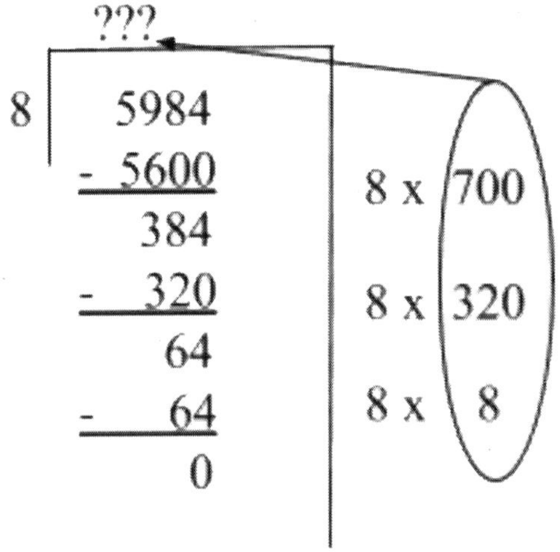

$$\begin{array}{r|r|l} & ??? & \\ \hline 8 & 5984 & \\ & \underline{-\ 5600} & 8 \times 700 \\ & 384 & \\ & \underline{-\ 320} & 8 \times 320 \\ & 64 & \\ & \underline{-\ 64} & 8 \times 8 \\ & 0 & \end{array}$$

Figure G.11.

partial quotient. A strategy used to divide where a student finds the quotients of part of a division problem and then adds all the quotients together to find the total quotient (e.g., 5984 ÷ 8 = 1028). (See figure G.11.)

partial sums. The addition of parts by place value, rather than using the traditional algorithm. With partial sums, students work on one place value at a time, and then add all the partial sums to find the total sum (e.g., 70 + 20 = 90, and 3 + 5 = 8, so 90 + 8 = 98). (See figure G.12.)

partition. To divide or split into parts that are of equal size (e.g., the whole rectangle has been partitioned into four equal-sized pieces. (See figure G.13.)

pictorial. Use of pictures often of objects to represent a particular quantity when solving problems e.g., the drawing below represents five cookies in a problem. (See figure G.14.)

place value. Place value is the basis of our entire number system. A place-value system is one in which the position of a digit in a number determines its value. In the standard system, called base ten, each place represents 10 times the value of the place to its right. (See figure G.15.)

product. The result of a multiplication problem (e.g., in the equation 2 × 4 = 8, 8 is the product).

quotient. The result of a division problem (e.g., in the equation 8 ÷ 4 = 2, 2 is the quotient).

representation. In this stage, the teacher transforms the concrete model into a representational (semi-concrete) level. In other words, this is the "seeing" stage, using images of objects to solve problems. This stage involves

$$\begin{array}{rcl} 73 & \rightarrow & 70 + 3 \\ \underline{+\,25} & \rightarrow & \underline{20 + 5} \\ 98 & \rightarrow & 90 + 8 \end{array}$$

Figure G.12.

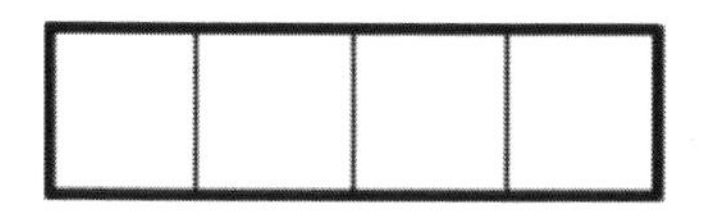

Figure G.13.

Figure G.14.

hund.	tens	ones
1	3	6

One hundred thirty-six

Figure G.15.

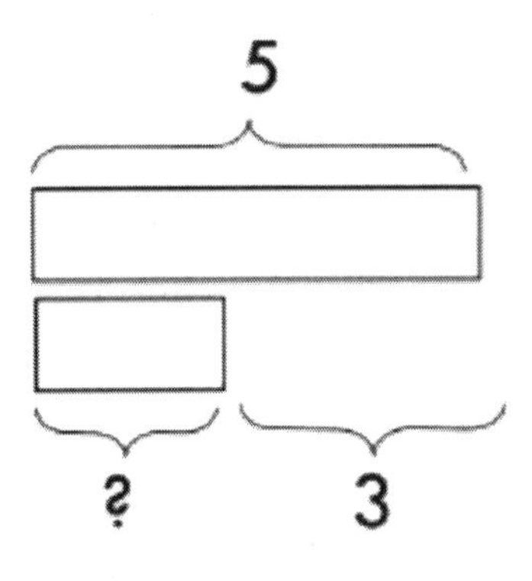

Figure G.16.

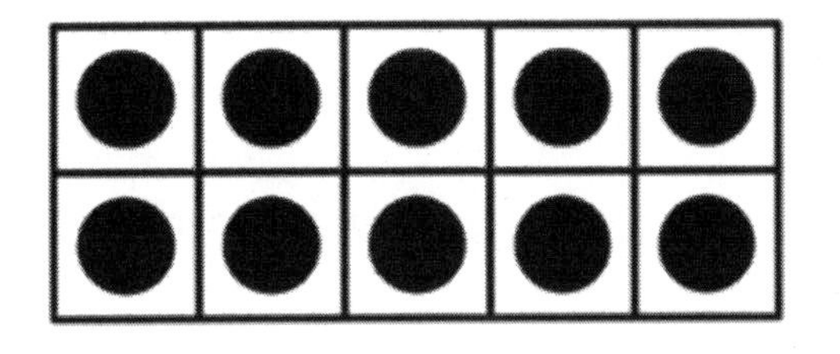

Figure G.17.

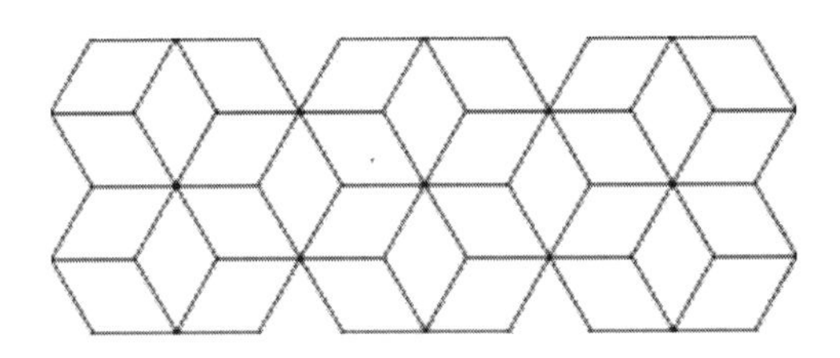

Figure G.18.

drawing pictures; using circles, dots, and tallies; or using stamps to imprint pictures for counting.

skip counting. Counting forward or backward by a number other than 1 (e.g., counting by twos, threes, fours, etc.).

solid. Three-dimensional (3-D) figure (e.g., cube, cone, pyramid, prism, etc.).

standard measurements. Units that are typically used within each measurement system to label the measurement of objects (e.g., inches, feet, meters).

subitize. To be able to see an arrangement of a quantity and instantly know how many objects are in the set. See chapter 3 for the difference between perceptual and conceptual subitizing (e.g., when looking at dice, you can easily identify the quantity without having to count the dots).

subtrahend. The quantity subtracted from another quantity (e.g., in the equation $4 - 3 = 1$, 3 is the subtrahend).

sum. The result of an addition problem (e.g., in the equation $2 + 3 = 5$, 5 is the sum).

tape diagrams. Tape diagrams are visual models that use rectangles to represent parts and wholes. Also known as bar models, these drawings illustrate number relationships and can sometimes be used to show comparisons. (See figure G.16.)

ten-frame. A ten-frame is a graphic tool that allows people to "see" the structure of numbers. Understanding that numbers are composed of tens and ones is an important foundational concept, setting the stage for work with larger numbers. (See figure G.17.)

tessellations. Arrangements of polygons in repeated patterns. (See figure G.18.)

unit fractions. Fractions where the numerator is 1 (e.g., $\frac{1}{2}$, $\frac{1}{4}$, $\frac{1}{8}$ . . .).

unitize. To be able to group individual units together to form a whole unit (e.g., 12 eggs is 1 dozen, where *dozen* represents the grouped unit and *eggs* represents individual units).

variables. Symbols that replace numbers (e.g., the letter x is commonly used to replace numbers, such as in the expression $2x + 4$).

vinculum. The line separating the numerator from the denominator of a fraction (e.g., $\frac{a}{b}$).

whole numbers. Numbers not written as fractions and do not include negative numbers (e.g., 0, 1, 2. . .).

zero pair. A pair of numbers whose sum is zero (e.g., $1 + (-1) = 0$).

References

Blackwell, L., Trzesniewski, K., & Dweck, C. (2007). Implicit theories of intelligence predict achievement across an adolescent transition: A longitudinal study and an intervention. *Child Development, 78*(1), 246–263.

Boaler, J. (2014). Research suggests that timed tests cause math anxiety. *Teaching Children Mathematics, 20*(8), 469–473.

Clements, D., & Sarama, J. (2014). *Learning and teaching early math: The learning trajectories approach* (2nd ed.). (*Studies in Mathematical Thinking and Learning Series.*) New York, NY: Routledge.

Dweck, C. (2006). *Mindset: The new psychology of success.* New York, NY: Random House.

Dyke, E., & Dyke, J. (2014). *10 things every parent should know about math.* Queen Creek, AZ: J&E Dyke Enterprises, LLC.

Fosnot, C. T., & Dolk, M. (2001). *Young mathematicians at work: Constructing multiplication and division.* Portsmouth, NH: Heinemann.

Frye, D., Baroody, A. J., Burchinal, M., Carver, S. M., Jordan, N. C., & McDowell, J. (2013). *Teaching math to young children: A practice guide* (NCEE 2014–2015). Washington, DC: National Center for Education Evaluation and Regional Assistance (NCEE), Institute of Education Sciences (IES), U.S. Department of Education. Retrieved from http://whatworks.ed.gov.

Gunderson, E. A., Gripshover, S. J., Romero, C., Dweck, C. S., Goldin-Meadow, S., & Levine, S. C. (2013). Parent praise to 1- to 3-year-olds predicts children's motivational frameworks 5 years later. *Child Development, 84*(5), 1526–1541.

Kac, M., Rota, G. C., & Schwartz, J. T. (1992). *Discrete thoughts: Essays on mathematics, science, and philosophy.* New York, NY: Springer Science+Business Media.

Lamon, S. J. (2012). *Teaching fractions and ratios for understanding: Essential content knowledge and instructional strategies for teachers* (3rd ed.). London: Routledge, Taylor & Francis Group.

Lesh, R., Landau, M., & Hamilton, E. (1983). Conceptual models in applied mathematical problem solving. In R. Lesh (Ed.), *The acquisition of mathematical concepts and processes*. New York, NY: Academic Press.

Levine, S. (2017). https://www.greatschools.org/gk/articles/early-math-equals-future-success/.

Libertus, M. E., Feigenson, L., & Halberda, J. (2011). Preschool acuity of the approximate number system correlates with school math ability. *Developmental Sciences, 14*(6), 1292–1300.

Merzenich, M. (2013). *Soft-wired: How the new science of brain plasticity can change your life* (2nd ed.). San Francisco, CA: Parnassus Publishing.

Miller, G. A. (1956). The magical number seven, plus or minus two: Some limits on our capacity for processing information. *Psychological Review, 63*(2), 81–97. [This work is one of the most highly cited papers in psychology.]

Moser, J., Schroder, H. S., Heeter, C., Moran, T. P., & Lee, Y. H. (2011). Mind your errors: Evidence for a neural mechanism linking growth mindset to adaptive post error adjustments. *Psychological Science, 22*(12), 1484–1489.

National Research Council. (2001). *Adding it up: Helping children learn mathematics*. J. Kilpatrick, J. Swafford, & B. Findell (Eds.). Mathematics Learning Study Committee, Center for Education, Division of Behavioral and Social Sciences and Education. Washington, DC: National Academy Press.

Richardson, K. (1999). *Developing number concepts. Book 2: Addition and subtraction*, p. 43. Lebanon, IN: Pearson Learning Group/Dale Seymour Publications.

Siegler, R. S., Duncan, G. J., Davis-Kean, P. E., Duckworth, K., Claessens, A., Engel, M., Susperreguy, M. I., & Chen, M. (2012). Early predictors of high school mathematics achievement. *Psychological Science, 23*(7), 691–697.

Sousa, D. (2015). *How the brain learns mathematics* (2nd ed.). Thousand Oaks, CA: Corwin.

Van de Walle, J., Karp, K., & Bay-Williams, J. (2017). *Elementary and middle school mathematics* (9th ed.). London, UK: Pearson Education.

Walker, T. M. No, parental involvement is not "overated." *NEA Today,* April 24, 2014.

World Economic Forum. (2016). The future of jobs: Employment, skills and workforce strategy for the fourth industrial revolution. Retrieved from http://www3.weforum.org/docs/WEF_Future_of_Jobs.pdf.

Wright, R. J., Martland, J., Stafford, A. K., & Stanger, G. (2006). *Teaching number: Advancing children's skills and strategies* (2nd edition). London: Sage.

About the Authors

Hilary Kreisberg is the director of the Center for Mathematics Achievement and an assistant professor of mathematics education at Lesley University in Cambridge, Massachusetts. She began her career as an elementary teacher and later became a K–5 math coach to be able to support other teachers in their understanding and teaching of mathematics.

As a member of several local, regional, and national mathematics teacher organizations, she became interested in educational leadership and was elected president of the Boston Area Mathematics Specialists, a professional development network for supervisors of mathematics. Hilary is a certified U.S. math recovery intervention specialist, a global math project ambassador, and a reviewer for National Council of Teachers of Mathematics (NCTM) journals, as well as a local, regional, and national speaker.

She holds a bachelor's degree in mathematics, a master's degree in teaching, and a doctorate in educational leadership and curriculum development. She is also endorsed to teach sheltered English immersion learners and holds both special education and mathematics licensure. For fun, Hilary likes to do Zumba and play chess. Find her on Twitter: @Dr_Kreisberg.

Matthew L. Beyranevand is the K−12 Mathematics Department coordinator for the Chelmsford, Massachusetts, public schools. He is an ambassador for the Global Math Project, supporter for the With Math I Can campaign, and member of the Massachusetts STEM Advisory Council. He also serves as an adjunct professor of mathematics and education at the University of Massachusetts at Lowell and Fitchburg State University. He is the author of the book *Teach Math Like This, Not Like That*, and his website is www.mathwithmatthew.com. Find him on Twitter: @Mathwithmatthew.